LEVEL C

Options
Publishing™

Table of Contents

Number Sense and Operations

Algebra

Geometry

Measurement

Data Analysis and Probability

Numbers to 100,000

Count the blocks. Then fill in the blanks.

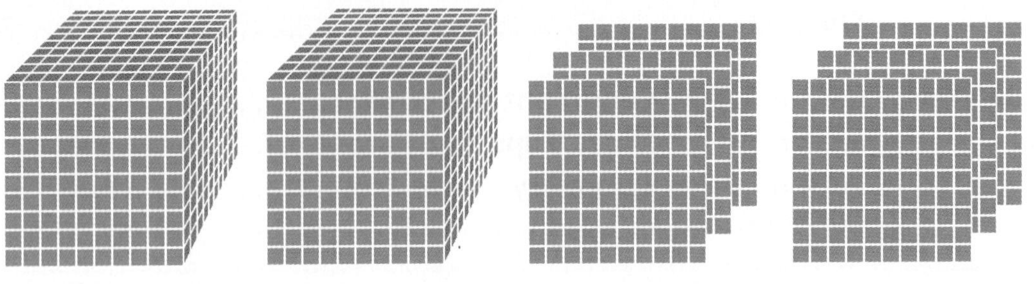

2 thousands, 6 hundreds, 0 tens, and 4 ones equal 2,604 in all.

Read the words. Write the number.

seventy-three thousand, eight hundred twenty-nine

73,829

Read the number. Write the words.

50,641 ______________________________

Count the blocks. Then fill in the blanks.

1

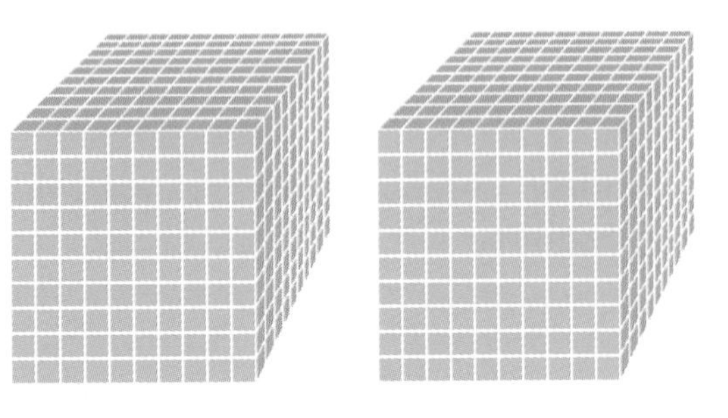

_______ thousands, _______ hundreds, _______ tens, and _______ ones equal _______________ in all.

Read the words. Write the number.

2 sixteen thousand, two hundred eighty ____________

3 four thousand, nine hundred thirty-three ____________

4 eighty-nine thousand, fifteen ____________

5 forty-five thousand, three hundred seven ____________

Read the number. Write the words.

6 67,149 ____________________________

____

7 5,302 ____________________________

____

8 19,999 ____________________________

____

9 80,880 ____________________________

____

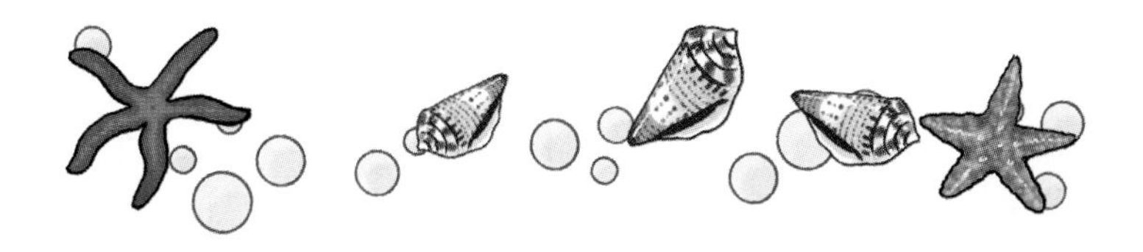

Place Value

▶ **Count the blocks. Then fill in the chart.**

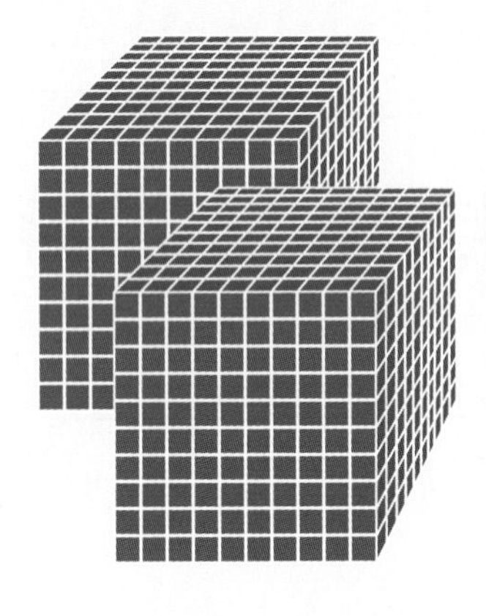

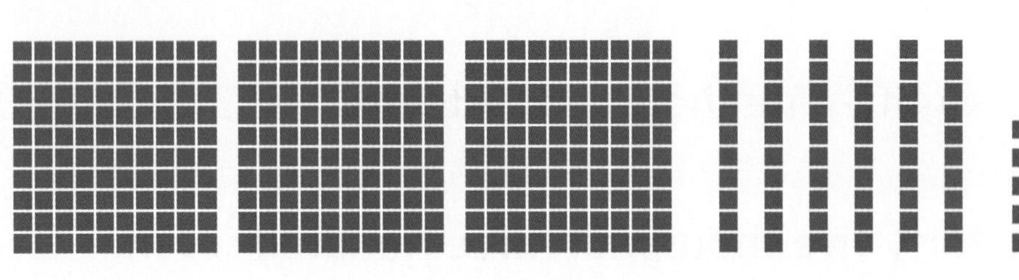

Thousands	Hundreds	Tens	Ones
2 ,	3	6	5

▶ **Fill in the chart.**

8,705

Thousands	Hundreds	Tens	Ones
____ ,	____	____	____

▶ **Write the number.**

____

Thousands	Hundreds	Tens	Ones
6 ,	3	1	9

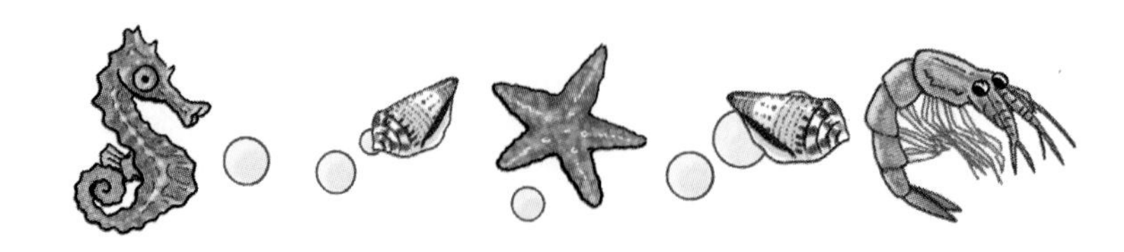

Count the blocks. Then fill in the chart.

1

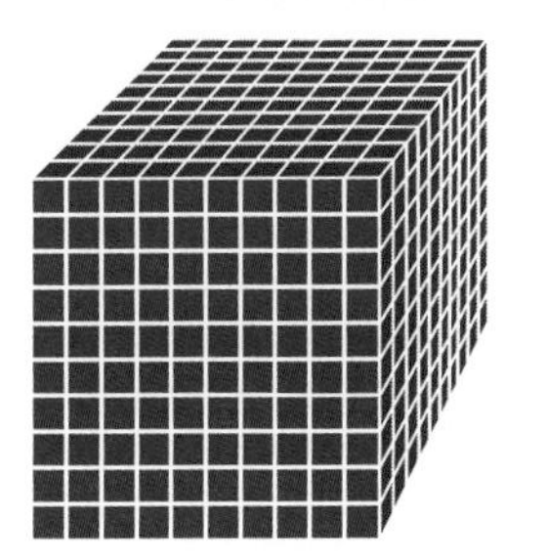

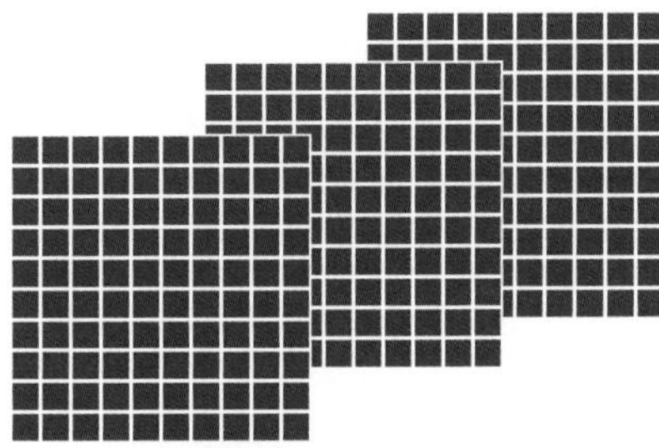

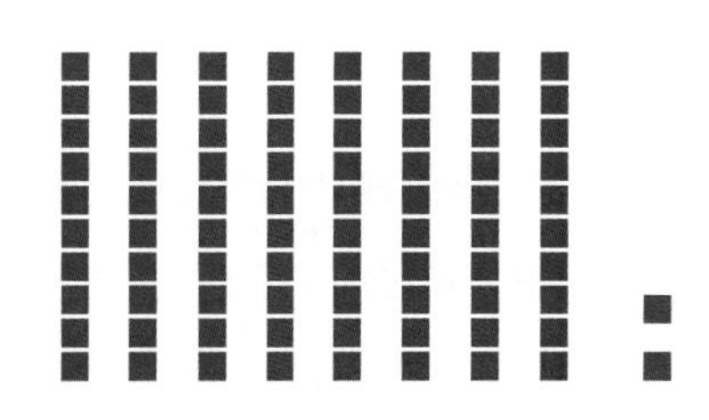

Thousands	Hundreds	Tens	Ones
____ ,	____	____	____

Fill in each chart.

2 9,191

Thousands	Hundreds	Tens	Ones
____ ,	____	____	____

3 5,442

Thousands	Hundreds	Tens	Ones
____ ,	____	____	____

Write each number.

4 ____________

Thousands	Hundreds	Tens	Ones
7 ,	1	5	0

5 ____________

Thousands	Hundreds	Tens	Ones
3 ,	9	7	6

Compare and Order Numbers

You can use a number line to compare and order numbers. The lesser number is always to the left of the greater number on a number line.

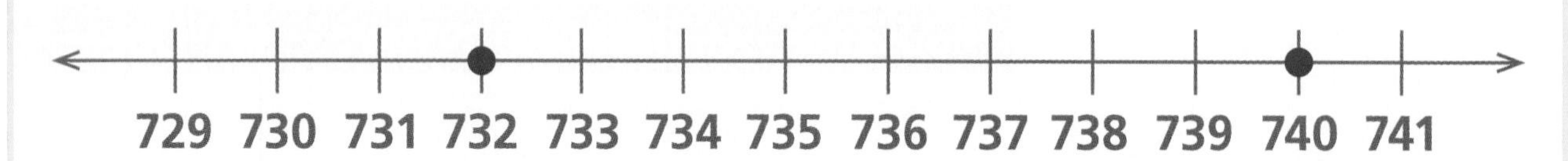

732 $<$ 740 — 732 is less than 740.

740 $>$ 732 — 740 is greater than 732.

Compare the numbers. Write $>$, $<$, or $=$ in the circle.

735 ◯ 730 — 735 is greater than 730.

Circle these numbers on the number line: 2,503 2,508 2,501.

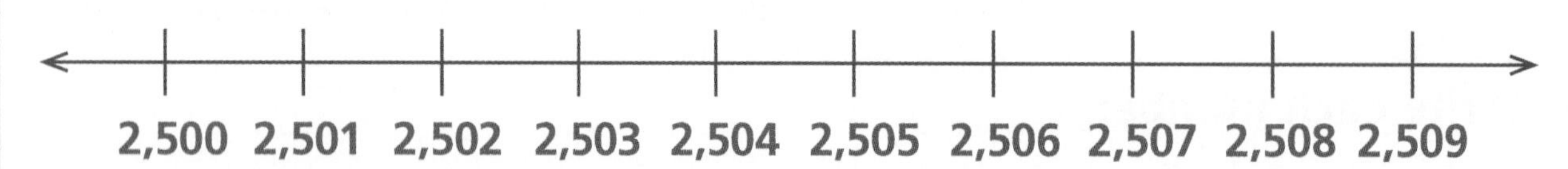

The number furthest to the left has the least value.
The number furthest to the right has the greatest value.

Now write the numbers in order from least to greatest.

________, ________, ________

Compare the numbers. Write $>$, $<$, or $=$ in the circle.

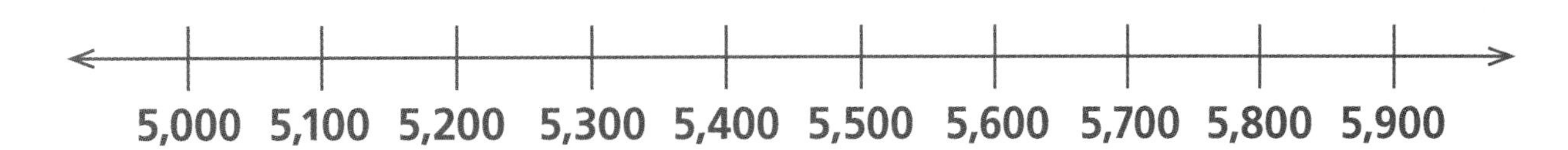

1 5,700 ◯ 5,800

2 5,000 ◯ 5,000

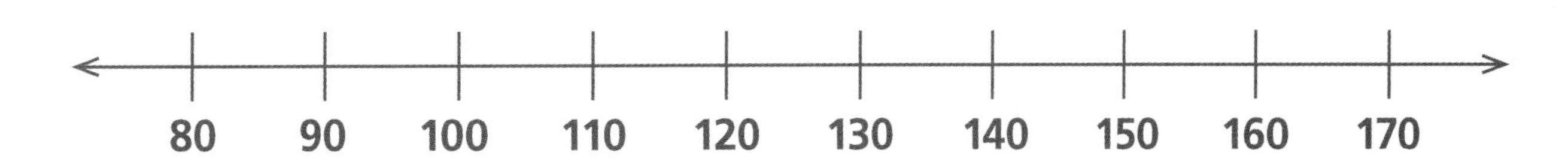

3 90 ◯ 160

4 150 ◯ 170

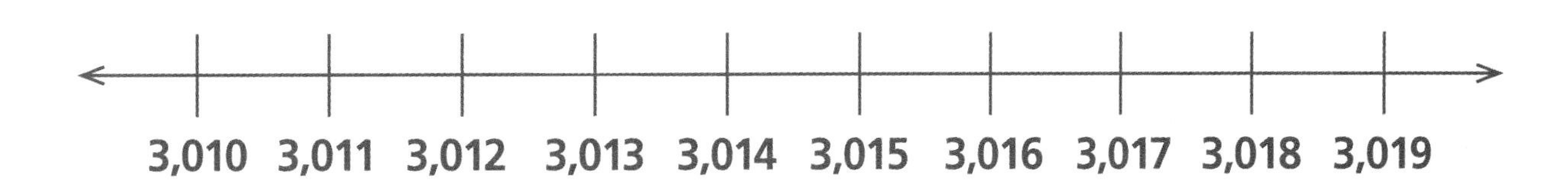

5 3,017 ◯ 3,018

6 3,012 ◯ 3,010

Use the number lines above to help you order the numbers from least to greatest.

7 5,400 5,100 5,200 ______, ______, ______

8 100 130 110 ______, ______, ______

9 3,019 3,010 3,020 ______, ______, ______

10 160 90 140 ______, ______, ______

Lesson

Rounding

You can use number lines to **round** numbers. If a number is halfway between two numbers, it rounds to the higher number.

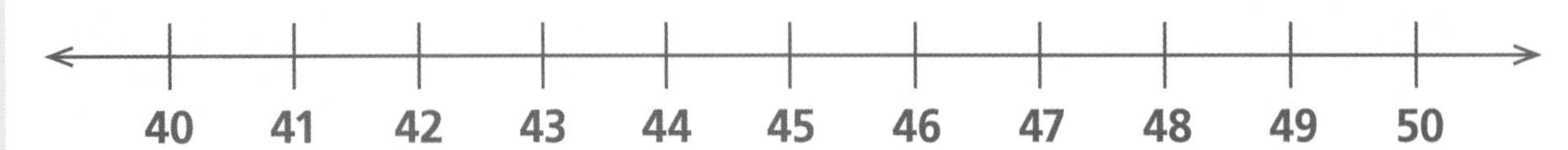

Rounded to the nearest ten,

43 rounds to 40.

45 rounds to 50.

46 rounds to ________.

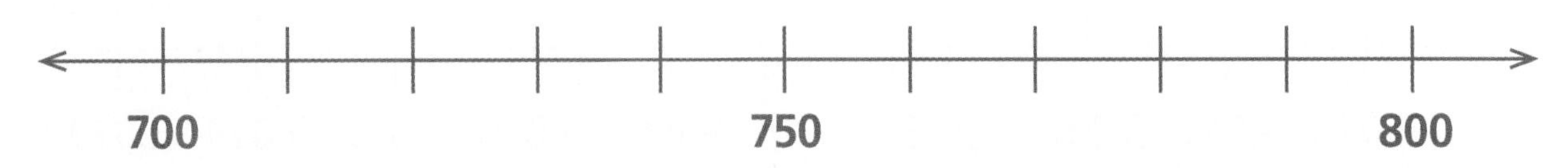

Rounded to the nearest hundred,

764 rounds to 800.

750 rounds to ________.

708 rounds to ________.

Use the number lines above to round each number.

1 To the nearest ten, 41 rounds to ________.

2 To the nearest hundred, 739 rounds to ________.

You can also round without using a number line. Let's use this example.

Round 137 to the nearest ten.

- Circle the number in the tens place. 137
- Underline the number to the right of the circled number. 137

  If the underlined number **is 5 or more**,
  add 1 to the circled number.
  If the underlined number is **less than 5**,
  leave the circled number the same, and
  replace the number to the right with a 0.

7 $>$ **5**, so to the nearest ten, 137 rounds to 140.

Round 1,568 to the nearest hundred.

Circle the ________ and underline the ________.

________ $>$ 5, so to the nearest hundred,

1,568 rounds to ________.

Round to the nearest ten.

3 761 ________

4 44 ________

5 185 ________

6 7,651 ________

Round to the nearest hundred.

7 285 ________

8 1,325 ________

9 542 ________

10 2,761 ________

Checkup

Count the blocks. Then fill in the blanks.

1 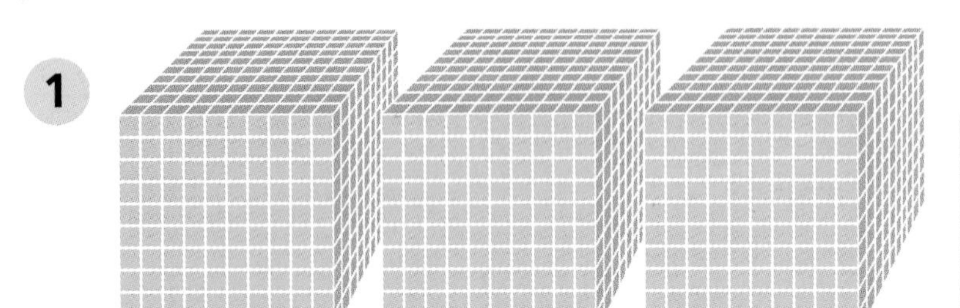

_______ thousands, _______ hundreds, _______ ten, and _______ ones

Read the words. Write the number.

2 nine thousand, seven hundred twenty-three ______________________

3 eighty thousand, five hundred twelve ______________________

Read the number. Write the words.

4 67,109 ____

____

Count the blocks. Then fill in the chart.

5 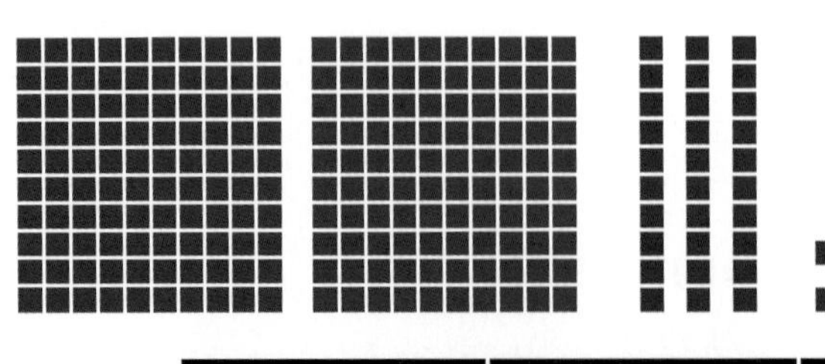

Thousands	Hundreds	Tens	Ones
______ ,	______	______	______

Fill in the chart.

6 8,705

Thousands	Hundreds	Tens	Ones
______ ,	______	______	______

Compare the numbers. Write $>$, $<$, or $=$ in the circle.

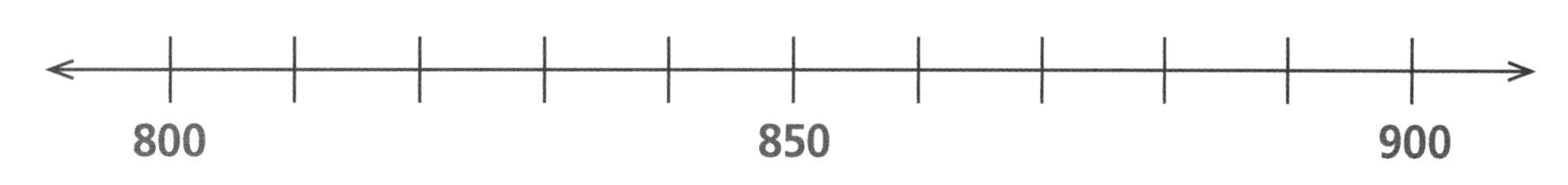

7 805 ◯ 820

8 880 ◯ 800

Order the numbers from least to greatest.

9 840 820 835 ______ , ______ , ______

10 889 819 890 ______ , ______ , ______

Use the number line to round the number.

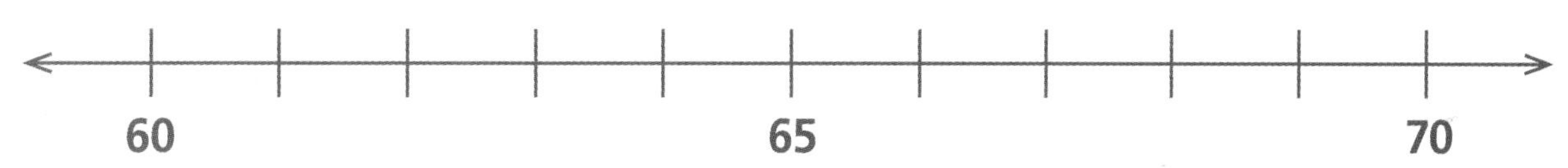

11 To the nearest ten, 65 rounds to ______.

Round each number.

12 Round 881 to the nearest hundred. ______

13 Round 3,153 to the nearest ten. ______

Lesson

Addition Without Regrouping

You can estimate the sum of large numbers. Start by rounding each addend to the greatest place.

Estimate the sum of 607 + 182. Begin by rounding each addend to the nearest hundred.

607 rounds to 600.

182 rounds to 200.

$$\begin{array}{r} 600 \\ +\ 200 \\ \hline 800 \end{array}$$

Now add 607 + 182. Compare the actual sum to the estimated sum.

- First, add the ones.

  7 + 2 = ________

- Then, add the tens.

  0 + 8 = ________

- Then, add the hundreds.

  6 + 1 = ________

	Hundreds	Tens	Ones
	6	0	7
+	1	8	2

607 + 182 = ________

Estimate:

$$\begin{array}{r} 6\,0\,0 \\ +\ 2\,0\,0 \\ \hline \end{array}$$

Actual:

$$\begin{array}{r} 6\,0\,7 \\ +\ 1\,8\,2 \\ \hline \end{array}$$

Is the estimated sum close to the actual sum? Circle your answer.

yes no

Estimate each sum. Then find the actual sum.

1 851 + 436 = ?

851 rounds to ______.

436 rounds to ______.

$$\begin{array}{r} \_\_\_\_ \\ +\ \_\_\_\_ \\ \hline \_\_\_\_ \end{array} \qquad \begin{array}{r} 851 \\ +\ 436 \\ \hline \end{array}$$

2 778 + 711 = ?

778 rounds to ______.

711 rounds to ______.

$$\begin{array}{r} \_\_\_\_ \\ +\ \_\_\_\_ \\ \hline \_\_\_\_ \end{array} \qquad \begin{array}{r} 778 \\ +\ 711 \\ \hline \end{array}$$

Find each sum.

3 105 + 681 = ?

	Hundreds	Tens	Ones
	1	0	5
+	6	8	1

4 281 + 615 = ?

	Hundreds	Tens	Ones
	2	8	1
+	6	1	5

5
$$\begin{array}{r} 567 \\ +\ 131 \\ \hline \end{array}$$

6
$$\begin{array}{r} 310 \\ +\ 52 \\ \hline \end{array}$$

7
$$\begin{array}{r} 164 \\ +\ 235 \\ \hline \end{array}$$

8
$$\begin{array}{r} 315 \\ +\ 144 \\ \hline \end{array}$$

9
$$\begin{array}{r} 418 \\ +\ 250 \\ \hline \end{array}$$

10
$$\begin{array}{r} 765 \\ +\ 234 \\ \hline \end{array}$$

Lesson 6 Addition with Regrouping

Find the sum of 145 + 87.

- First, add the ones.

  5 + 7 = 12

- Regroup 12 as 1 ten and 2 ones.
- Write a 2 in the ones place and a 1 above the tens digits.

	Hundreds	Tens	Ones
		1	
	1	4	5
+		8	7
			2

- Next, add the tens.

  1 + 4 + 8 = 13

- Regroup 13 tens as 1 hundred and 3 tens.
- Write a 3 in the tens place and a 1 above the hundreds digit.

	Hundreds	Tens	Ones
	1	1	
	1	4	5
+		8	7
		3	2

- Then, add the hundreds.

  1 + 1 = 2

- Write a 2 in the hundreds place.

	Hundreds	Tens	Ones
	1	1	
	1	4	5
+		8	7
	2	3	2

145 + 87 = ____________

Find each sum.

1. 735 + 87 = ?

	Hundreds	Tens	Ones
	7	3	5
+		8	7

2. 202 + 498 = ?

	Hundreds	Tens	Ones
	2	0	2
+	4	9	8

3. $\begin{array}{r} 599 \\ +\ \ 62 \\ \hline \end{array}$

4. $\begin{array}{r} 708 \\ +295 \\ \hline \end{array}$

5. $\begin{array}{r} 786 \\ +366 \\ \hline \end{array}$

6. $\begin{array}{r} 167 \\ +854 \\ \hline \end{array}$

7. $\begin{array}{r} 53 \\ +\ \ 98 \\ \hline \end{array}$

8. $\begin{array}{r} 457 \\ +564 \\ \hline \end{array}$

9. $\begin{array}{r} 908 \\ +\ \ 92 \\ \hline \end{array}$

10. $\begin{array}{r} 303 \\ +798 \\ \hline \end{array}$

11. $\begin{array}{r} 121 \\ +189 \\ \hline \end{array}$

12. $\begin{array}{r} 239 \\ +175 \\ \hline \end{array}$

13. $\begin{array}{r} 482 \\ +119 \\ \hline \end{array}$

Lesson

Properties of Addition

The Commutative Property of Addition says the order in which you add does not change the sum.

Examples:

$7 + 8 = 8 + 7$
$15 = 15$

$3 + 9 = 9 + 3$
$12 = 12$

Use the Commutative Property of Addition to rewrite the sentence.

$482 + 97 =$ ______________________

The Associative Property of Addition says that the way you group numbers does not change the sum.

Examples:

$2 + (11 + 9) = 2 + (9 + 11)$
$2 + 20 = 2 + 20$
$22 = 22$

$(3 + 4) + 5 = 3 + (4 + 5)$
$7 + 5 = 3 + 9$
$12 = 12$

Use the Associative Property of Addition to rewrite the sentence.

$59 + (27 + 63) =$ ______________________

Use the properties of addition to find the missing numbers.

$79 + 652 = 652 +$ ___________

$(34 + 6) + 10 = 34 + ($ ___________ $+ 10)$

▶ **Use the Commutative Property to rewrite each number sentence.**

1 9 + 11 = ____________________

2 75 + 51 = ____________________

3 913 + 526 = ____________________

4 547 + 495 = ____________________

▶ **Use the Associative Property to rewrite each number sentence.**

5 (8 + 7) + 15 = ____________________

6 19 + (11 + 13) = ____________________

7 27 + (41 + 9) = ____________________

8 (31 + 24) + 16 = ____________________

▶ **Use an addition property to find the missing numbers.**

9 66 + 18 = 18 + __________

10 51 + (43 + 17) = 51 + (__________ + 17)

11 31 + __________ = 9 + 31

12 (30 + 14) + 2 = __________ + (14 + 2)

Lesson

Subtraction Without Regrouping

You can estimate the difference of large numbers. Start by rounding each number to the greatest place.

Estimate 486 − 356. Begin by rounding each number to the nearest hundred.

486 rounds to 500.

356 rounds to 400.

$$\begin{array}{r} 500 \\ -\ 400 \\ \hline 100 \end{array}$$

Now subtract 356 from 486. Compare the actual difference to the estimated difference.

- First, subtract the ones.

  6 − 6 = ________

- Next, subtract the tens.

  8 − 5 = ________

- Then, subtract the hundreds.

  4 − 3 = ________

	Hundreds	Tens	Ones
	4	8	6
−	3	5	6

486 − 356 = ________

Estimate: $\begin{array}{r} 500 \\ -\ 400 \\ \hline \end{array}$

Actual: $\begin{array}{r} 486 \\ -\ 356 \\ \hline \end{array}$

Is the estimated difference close to the actual difference?

Circle your answer. yes no

Estimate each difference. Then find the actual difference.

1 59 − 28 = ?

59 rounds to ______.

28 rounds to ______.

$$\begin{array}{r} \_\_\_ \\ - \_\_\_ \\ \hline \_\_\_ \end{array} \qquad \begin{array}{r} 59 \\ - 28 \\ \hline \end{array}$$

2 447 − 220 = ?

447 rounds to ______.

220 rounds to ______.

$$\begin{array}{r} \_\_\_ \\ - \_\_\_ \\ \hline \_\_\_ \end{array} \qquad \begin{array}{r} 447 \\ - 220 \\ \hline \end{array}$$

Find each difference.

3 688 − 413 = ?

	Hundreds	Tens	Ones
	6	8	8
−	4	1	3

4 637 − 34 = ?

	Hundreds	Tens	Ones
	6	3	7
−		3	4

5 $\begin{array}{r} 349 \\ - 231 \\ \hline \end{array}$

6 $\begin{array}{r} 181 \\ - 30 \\ \hline \end{array}$

7 $\begin{array}{r} 767 \\ - 503 \\ \hline \end{array}$

8 $\begin{array}{r} 381 \\ - 271 \\ \hline \end{array}$

9 $\begin{array}{r} 839 \\ - 534 \\ \hline \end{array}$

10 $\begin{array}{r} 275 \\ - 163 \\ \hline \end{array}$

Lesson

Subtraction with Regrouping

Find the difference between 253 and 96.

- Regroup one ten as 10 ones.

  10 + 3 = 13

- Now, subtract the ones.

  13 − 6 = 7

- Write 7 in the ones place.

	Hundreds	Tens	Ones
		4	13
	2	~~5~~	~~3~~
−		9	6
			7

- Regroup one hundred as 10 tens.

  10 + 4 = 14

- Now, subtract the tens.

  14 − 9 = 5

- Write 5 in the tens place.

	Hundreds	Tens	Ones
	1	14	13
	~~2~~	~~5~~	~~3~~
−		9	6
		5	7

- Then, subtract the hundreds.

  1 − 0 = 1

- Write a 1 in the hundreds place.

	Hundreds	Tens	Ones
	1	14	13
	~~2~~	~~5~~	~~3~~
−		9	6
	1	5	7

253 − 96 = ___________

Regroup to find each difference.

1 413 − 246 = ?

	Hundreds	Tens	Ones
	4	1	3
−	2	4	6

2 157 − 88 = ?

	Hundreds	Tens	Ones
	1	5	7
−		8	8

3 $\begin{array}{r} 816 \\ -207 \\ \hline \end{array}$

4 $\begin{array}{r} 53 \\ -\ 36 \\ \hline \end{array}$

5 $\begin{array}{r} 708 \\ -397 \\ \hline \end{array}$

6 $\begin{array}{r} 117 \\ -109 \\ \hline \end{array}$

7 $\begin{array}{r} 60 \\ -\ 43 \\ \hline \end{array}$

8 $\begin{array}{r} 815 \\ -128 \\ \hline \end{array}$

9 $\begin{array}{r} 93 \\ -\ 89 \\ \hline \end{array}$

10 $\begin{array}{r} 503 \\ -346 \\ \hline \end{array}$

11 $\begin{array}{r} 652 \\ -289 \\ \hline \end{array}$

12 $\begin{array}{r} 132 \\ -\ 86 \\ \hline \end{array}$

13 $\begin{array}{r} 301 \\ -123 \\ \hline \end{array}$

Checkup

Estimate the sum. Then find the actual sum.

1. 941 + 256 = ?

941 rounds to ______.

256 rounds to ______.

```
   ____        9 4 1
 + ____      + 2 5 6
 ------      -------
   ____        _____
```

Find each sum.

2. 213 + 575 = ?

	Hundreds	Tens	Ones
	2	1	3
+	5	7	5

3. 305 + 284 = ?

	Hundreds	Tens	Ones
	3	0	5
+	2	8	4

Regroup to find each sum.

4. 818 + 95 = ?

	Hundreds	Tens	Ones
	8	1	8
+		9	5

5. 278 + 66 = ?

	Hundreds	Tens	Ones
	2	7	8
+		6	6

▶ **Use the Commutative Property to rewrite the number sentence.**

6 415 + 185 = ________________

▶ **Use the Associative Property to rewrite the number sentence.**

7 63 + (89 + 74) = ________________

▶ **Estimate the difference. Then find the actual difference.**

8 135 rounds to ______.

24 rounds to ______.

	Estimate	Actual
	______	1 3 5
−	______	− 2 4
	______	______

▶ **Find each difference.**

9 381 − 271 = ?

	Hundreds	Tens	Ones
	3	8	1
−	2	7	1

10 637 − 534 = ?

	Hundreds	Tens	Ones
	6	3	7
−	5	3	4

▶ **Regroup to find each difference.**

11 343 − 187 = ?

	Hundreds	Tens	Ones
	3	4	3
−	1	8	7

12 801 − 271 = ?

	Hundreds	Tens	Ones
	8	0	1
−	2	7	1

Explore Multiplication

Multiplication is the same as **repeated addition**. In a multiplication sentence, the answer is the **product**.

▶ **Find the total number of apples.**

You can skip count: 3, 6, 9, 12.

You can use a hundred chart:

1	2	(3)	4	5	(6)	7	8	(9)	10
11	(12)	13	14	15	16	17	18	19	20
21	22	23	24	25	26	27	28	29	30
31	32	33	34	35	36	37	38	39	40
41	42	43	44	45	46	47	48	49	50
51	52	53	54	55	56	57	58	59	60
61	62	63	64	65	66	67	68	69	70
71	72	73	74	75	76	77	78	79	80
81	82	83	84	85	86	87	88	89	90
91	92	93	94	95	96	97	98	99	100

You can use addition:

$3 + 3 + 3 + 3 = 12$

You can multiply: There are 4 groups of 3 apples each.

4 groups of 3 = $4 \times 3 =$ ______

Find the total number of flowers.

1

Skip count: ______ , ______ , ______ , ______ , ______

Hundred chart: ______ , ______ , ______ , ______ , ______

Add: ______ + ______ + ______ + ______ + ______ = ______

Multiply: ______ groups × ______ flowers in each group =

______ flowers

Fill in the blanks to solve each problem.

2

  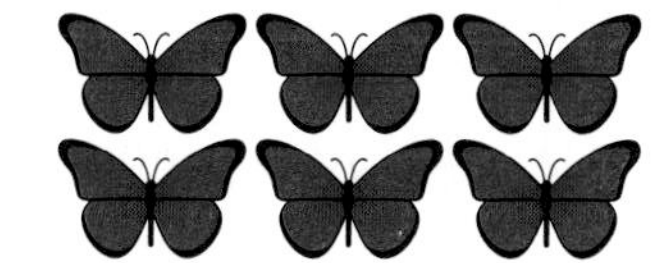

6 + 6 + 6 = ______

______ groups × ______ in each group = ______

3

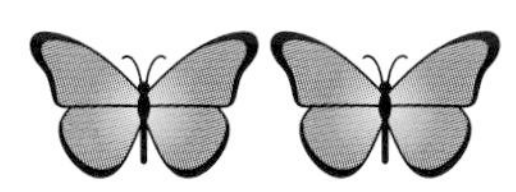 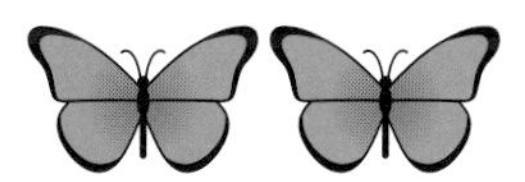 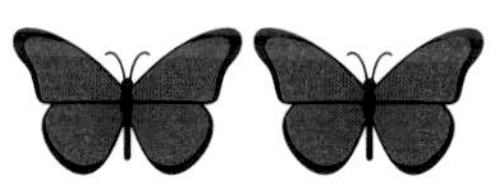 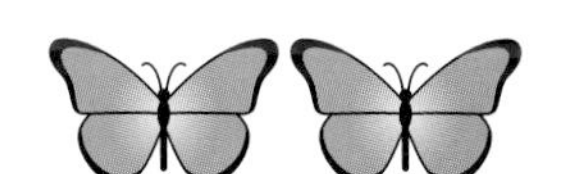

______ + ______ + ______ + ______ = ______

4 groups × 2 in each group = ______

4

______ + ______ + ______ = ______

______ groups × ______ in each group = ______

Arrays

An **array** is a group of objects shown in rows and columns.

The flags in this set are shown in 2 **rows** and 5 **columns**.

You can use an array to write a multiplication sentence.

$$2 \times 5 = 10$$

number of **rows** (number of sets) — 2

number of **columns** (number of items in each set) — 5

Write a multiplication sentence for the array.

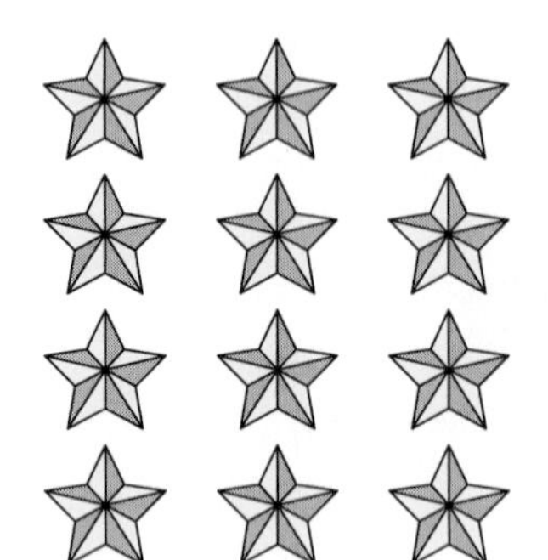

There are 4 sets.

There are 3 stars in each set.

4 × 3 = ______

Write a multiplication sentence for each array.

1

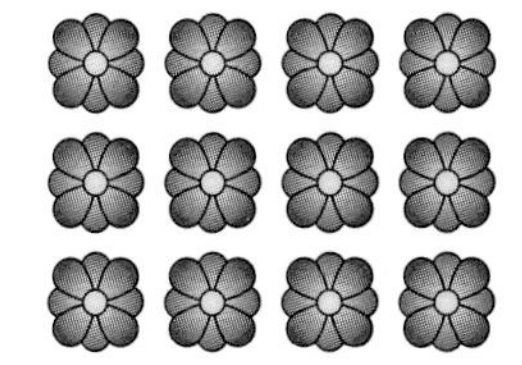

____ × ____ = ____

2

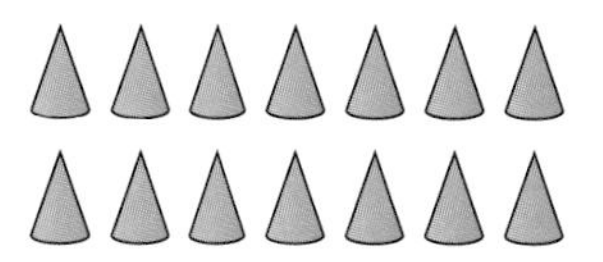

____ × ____ = ____

3

____ × ____ = ____

4

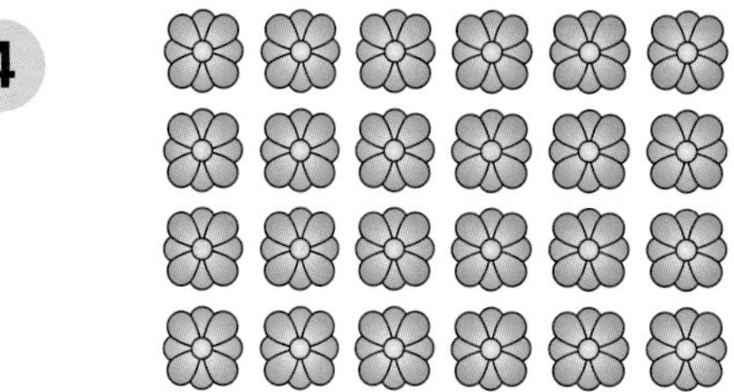

____ × ____ = ____

5

____ × ____ = ____

6

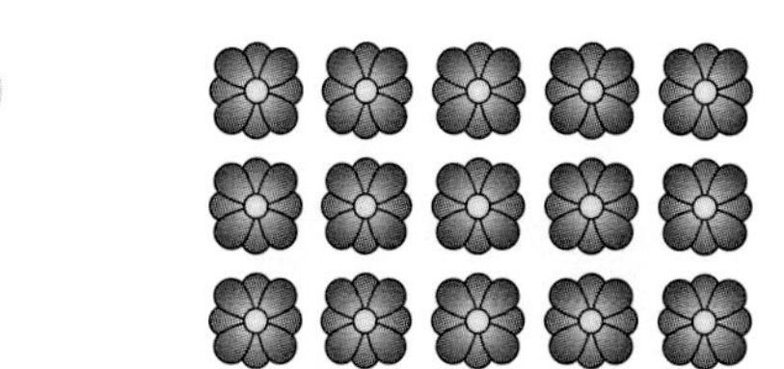

____ × ____ = ____

7

____ × ____ = ____

8

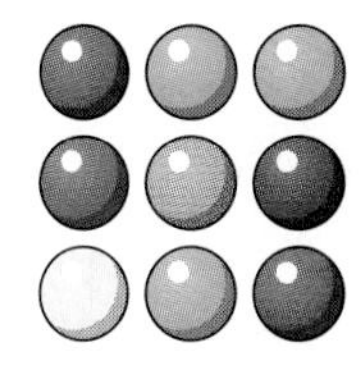

____ × ____ = ____

9

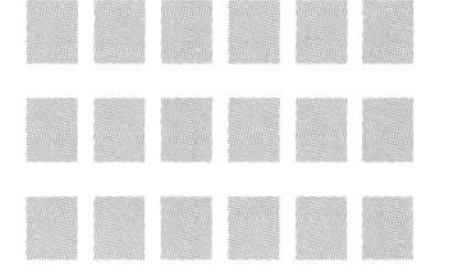

____ × ____ = ____

10

____ × ____ = ____

Lesson 12 Using a Multiplication Table

You can use a multiplication table to solve and check multiplication facts.

Example: $2 \times 3 = 6$ ← product

↑ ↑ factors

Find 2 in the first column. Find 3 in the first row. Use your finger to find the number where the column and row meet. This number is the product.

×	0	1	2	3	4	5	6	7	8	9	10
0	0	0	0	0	0	0	0	0	0	0	0
1	0	1	2	3	4	5	6	7	8	9	10
2	0	2	4	6	8	10	12	14	16	18	20
3	0	3	6	9	12	15	18	21	24	27	30
4	0	4	8	12	16	20	24	28	32	36	40
5	0	5	10	15	20	25	30	35	40	45	50
6	0	6	12	18	24	30	36	42	48	54	60
7	0	7	14	21	28	35	42	49	56	63	70
8	0	8	16	24	32	40	48	56	64	72	80
9	0	9	18	27	36	45	54	63	72	81	90
10	0	10	20	30	40	50	60	70	80	90	100

▶ **Use the table to find the products.**

A. $3 \times 4 =$ ______

B. $7 \times 6 =$ ______

Use the multiplication table to find the products.

1. $2 \times 6 =$ ______
2. $3 \times 7 =$ ______
3. $5 \times 4 =$ ______
4. $9 \times 1 =$ ______
5. $2 \times 5 =$ ______
6. $0 \times 8 =$ ______
7. $3 \times 9 =$ ______
8. $9 \times 8 =$ ______
9. $9 \times 5 =$ ______
10. $10 \times 10 =$ ______

Find each product. Then find the matching product below. Write the letter of the matching product on the line.

______ 11. $6 \times 8 =$ ______

______ 12. $8 \times 8 =$ ______

______ 13. $5 \times 7 =$ ______

______ 14. $3 \times 5 =$ ______

______ 15. $4 \times 9 =$ ______

______ 16. $9 \times 9 =$ ______

______ 17. $6 \times 4 =$ ______

______ 18. $6 \times 9 =$ ______

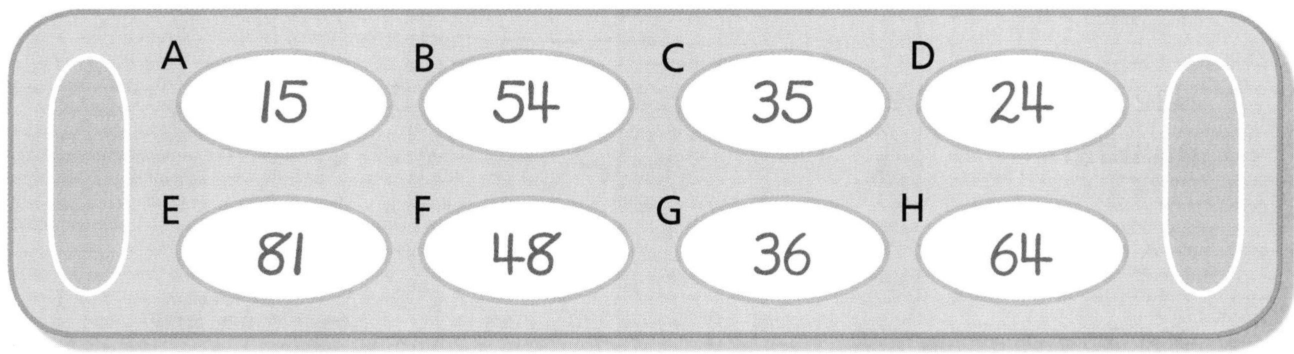

Properties of Multiplication

The **Commutative Property of Multiplication** says that you can change the order of the numbers in a multiplication sentence without changing the answer.

Examples: $8 \times 3 = 24$ $3 \times 8 = 24$

▶ **Use the Commutative Property of Multiplication to rewrite the sentence.**

$6 \times 12 =$ ______ $\times$ ______

The **Associative Property of Multiplication** says that you can move the parentheses in a multiplication sentence without changing the answer.

Examples: $(3 \times 1) \times 2 = 6$ $3 \times (1 \times 2) = 6$

$3 \times 2 = 6$ $3 \times 2 = 6$

▶ **Use the Associative Property of Multiplication to rewrite the sentence.**

$4 \times (1 \times 2) =$ ____________________

$4 \times 2 =$ ____________________

▶ **Use the properties of multiplication to find the missing numbers.**

$2 \times (6 \times 2) = (2 \times 6) \times$ ______

$9 \times 3 =$ ______ $\times 9$

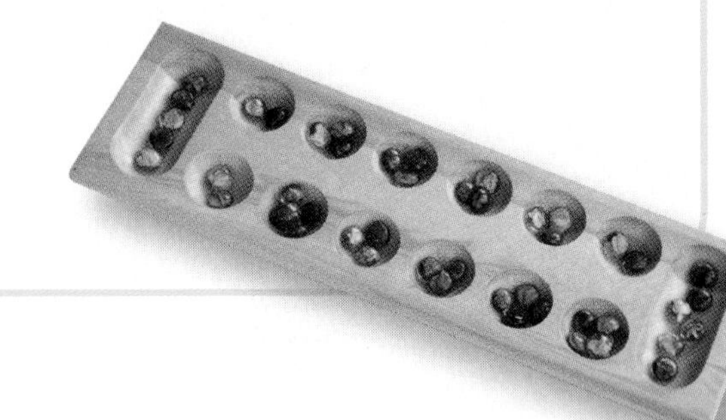

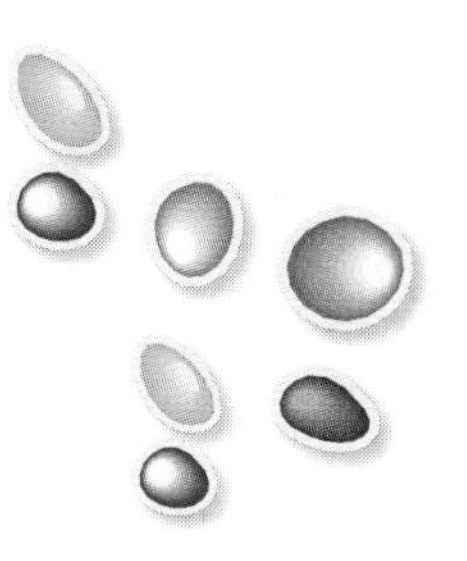

Use the Commutative Property to rewrite each multiplication sentence.

1. $4 \times 3 =$ ________________
2. $6 \times 5 =$ ________________
3. $8 \times 4 =$ ________________
4. $3 \times 9 =$ ________________

Use the Associative Property to rewrite each sentence.

5. $(2 \times 2) \times 3 =$ ________________
6. $1 \times (3 \times 5) =$ ________________
7. $(2 \times 3) \times 3 =$ ________________
8. $2 \times (2 \times 4) =$ ________________

Use the properties of multiplication to find the missing numbers.

9. $5 \times 7 = 7 \times$ ______
10. $3 \times (2 \times 4) = (3 \times$ ______ $) \times 4$
11. $8 \times$ ______ $= 6 \times 8$
12. (______ $\times 6) \times 3 = 1 \times (6 \times 3)$

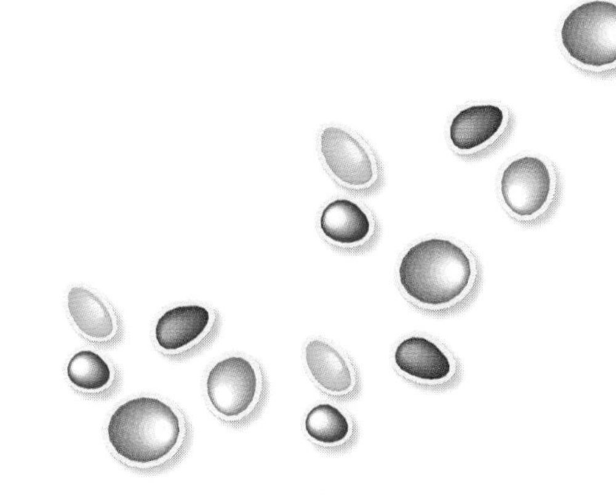

Multiplying by Regrouping

Sometimes you need to regroup to multiply numbers.

Find the product. 223 × 5 = ?

- Multiply the ones.
  5 × 3 = 15
- Regroup 15 as 1 ten and 5 ones.
- Write 5 in the ones place.

	Th	H	T	O
			1	
		2	2	3
×				5
	,			5

- Multiply the tens.
  5 × 2 = 10
- Add all the tens.
  10 + 1 = 11
- Regroup 11 tens as 1 hundred and 1 ten.
- Write 1 in the tens place.

	Th	H	T	O
		1	1	
		2	2	3
×				5
	,		1	5

- Multiply the hundreds.
  5 × 2 = 10
- Add all the hundreds.
  10 + 1 = 11
- Write a 1 in the hundreds place and 1 in the thousands place.

223 × 5 = ______

	Th	H	T	O
		1	1	
		2	2	3
×				5
	1 ,	1	1	5

Find each product.

1. $39 \times 2 = ?$

	H	T	O
		3	9
×			2

2. $47 \times 7 = ?$

	H	T	O
		4	7
×			7

3. $908 \times 1 = ?$

	H	T	O
	9	0	8
×			1

4. $265 \times 4 = ?$

	Th	H	T	O
		2	6	5
×				4
	,			

5. $\begin{array}{r} 346 \\ \times \quad 8 \\ \hline \end{array}$

6. $\begin{array}{r} 188 \\ \times \quad 4 \\ \hline \end{array}$

7. $\begin{array}{r} 913 \\ \times \quad 5 \\ \hline \end{array}$

8. $\begin{array}{r} 734 \\ \times \quad 2 \\ \hline \end{array}$

9. $\begin{array}{r} 280 \\ \times \quad 6 \\ \hline \end{array}$

10. $\begin{array}{r} 121 \\ \times \quad 9 \\ \hline \end{array}$

Checkup

Find the total number of cows.

1

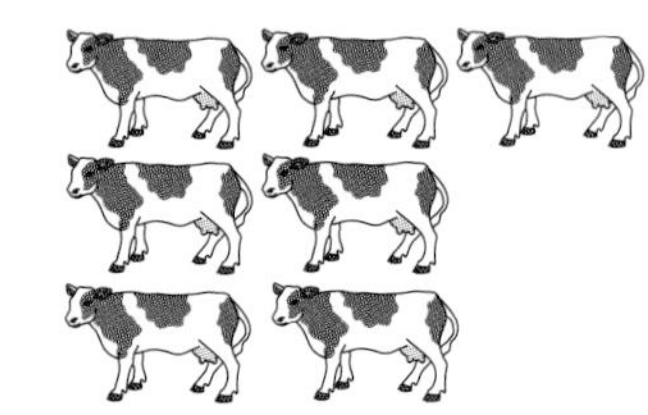

Skip count: ______ , ______ , ______

Add: ______ + ______ + ______ = ______

Multiply: ______ groups × ______ cows in each group =

______ cows

Write a multiplication sentence for each array.

2

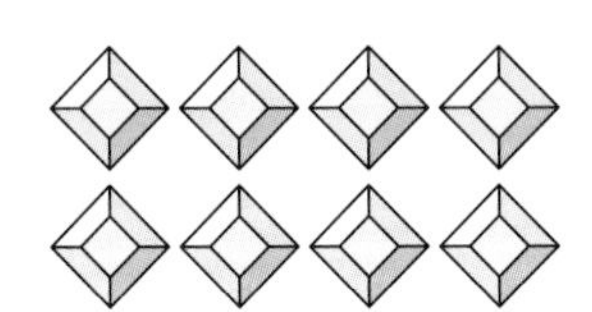

There are ______ sets.

There are ______ diamonds in each set.

______ × ______ = ______

3

______ × ______ = ______

Use the properties of multiplication to find the missing numbers.

4 5 × (3 × 7) = (5 × ______) × 7

5 9 × 6 = 6 × ______

▶ Use the multiplication table to find each product.

×	0	1	2	3	4	5	6	7	8	9	10
0	0	0	0	0	0	0	0	0	0	0	0
1	0	1	2	3	4	5	6	7	8	9	10
2	0	2	4	6	8	10	12	14	16	18	20
3	0	3	6	9	12	15	18	21	24	27	30
4	0	4	8	12	16	20	24	28	32	36	40
5	0	5	10	15	20	25	30	35	40	45	50
6	0	6	12	18	24	30	36	42	48	54	60
7	0	7	14	21	28	35	42	49	56	63	70
8	0	8	16	24	32	40	48	56	64	72	80
9	0	9	18	27	36	45	54	63	72	81	90
10	0	10	20	30	40	50	60	70	80	90	100

6 $8 \times 8 =$ ______

7 $6 \times 7 =$ ______

8 $4 \times 9 =$ ______

▶ Find each product.

9 $304 \times 3 = ?$

	H	T	O
	3	0	4
×			3

10 $99 \times 2 = ?$

	H	T	O
		9	9
×			2

11
$$\begin{array}{r} 624 \\ \times \quad 2 \\ \hline \end{array}$$

12
$$\begin{array}{r} 204 \\ \times \quad 7 \\ \hline \end{array}$$

13
$$\begin{array}{r} 515 \\ \times \quad 3 \\ \hline \end{array}$$

Division as Repeated Subtraction

Division is the same as **repeated subtraction**. In a division sentence, the answer is the **quotient**.

Use repeated subtraction to find the quotient. 8 ÷ 2 = ?

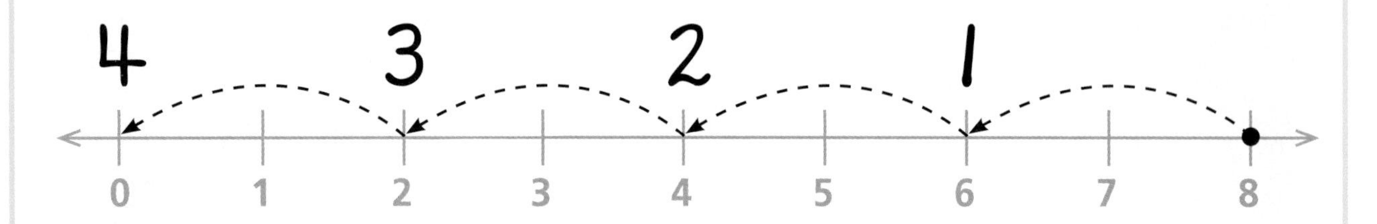

Start at 8. Go back by 2s until you are at 0.

You subtracted 4 times. **8 ÷ 2** = _______

Use repeated subtraction to find each quotient.

1 15 ÷ 3 = _______

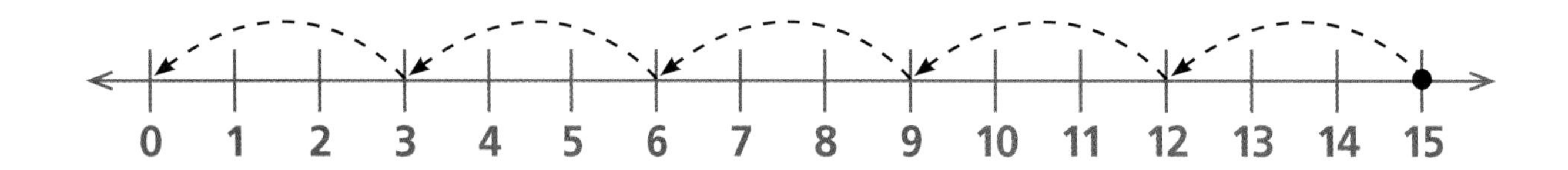

2 14 ÷ 2 = _______

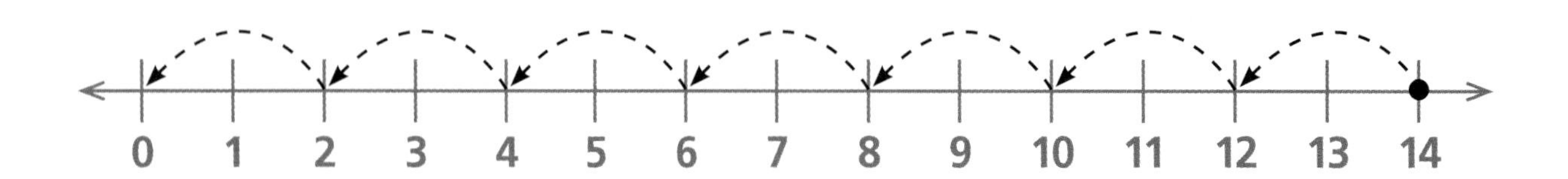

3 $25 \div 5 =$ ______

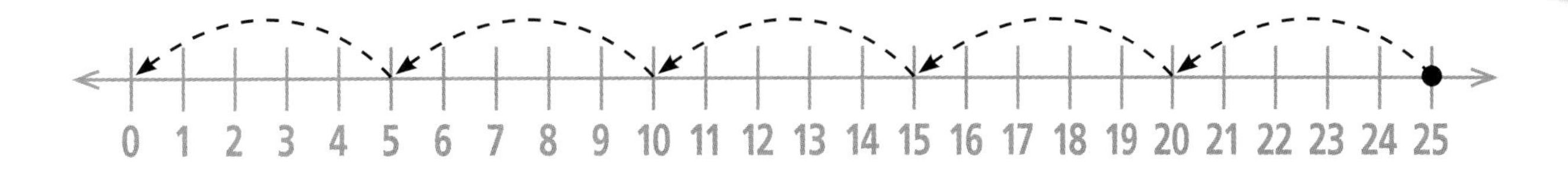

4 $12 \div 3 =$ ______

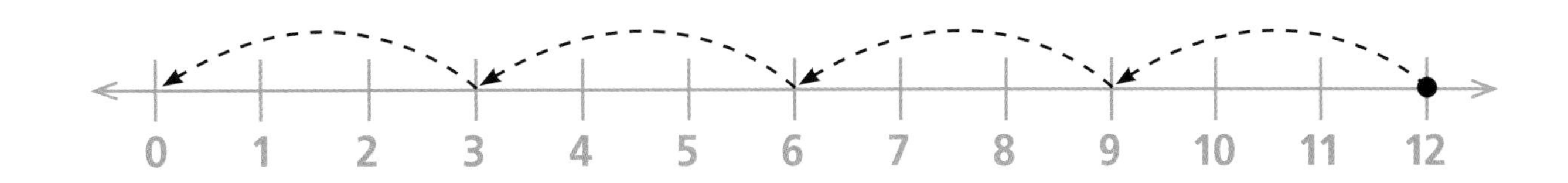

5 $30 \div 6 =$ ______

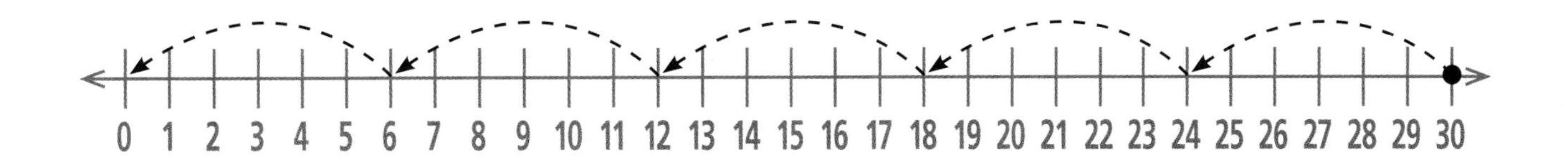

6 $16 \div 4 =$ ______

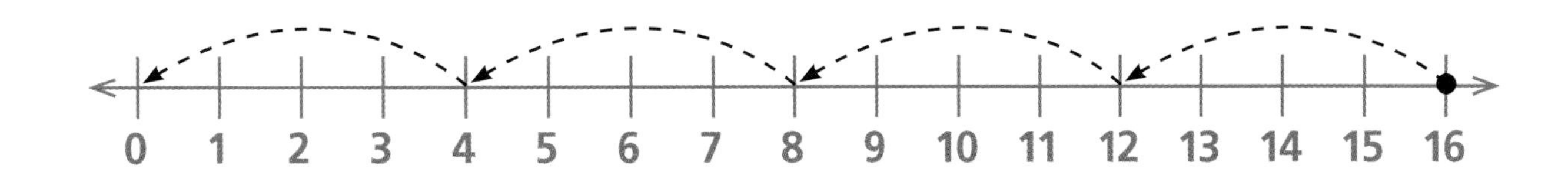

7 $10 \div 2 =$ ______

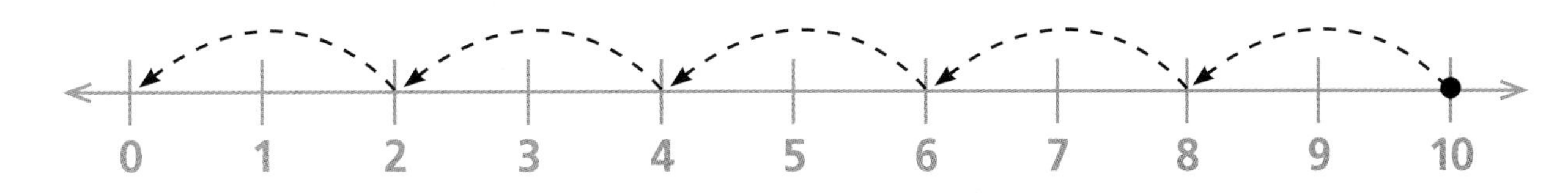

Lesson 16

Divide into Equal Groups

Division is the same as making equal groups.

Draw circles to make equal groups. Then fill in the blanks.

$12 \div 3 = ?$

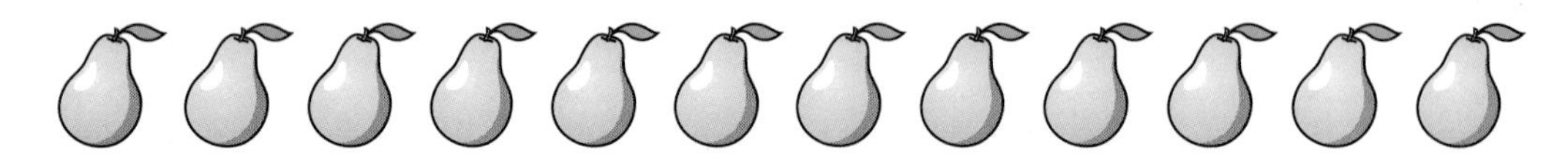

Draw a circle around each group of 3 pears.

There are ______ equal groups of 3. $12 \div 3 =$ ______

Draw circles to make equal groups. Then fill in the blanks.

1 $12 \div 4 = ?$

There are ______ equal groups of 4. $12 \div 4 =$ ______

2 $9 \div 3 = ?$

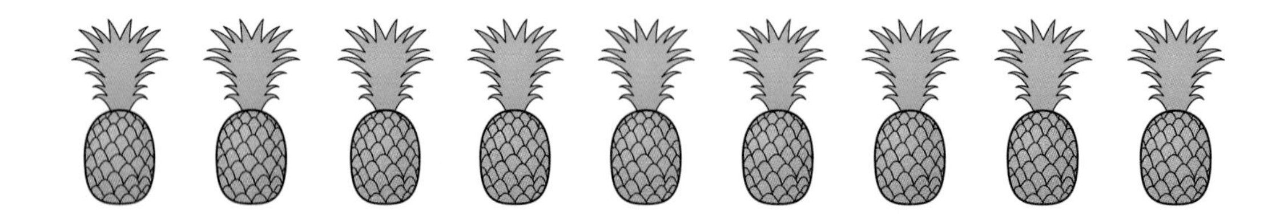

There are ______ equal groups of 3. $9 \div 3 =$ ______

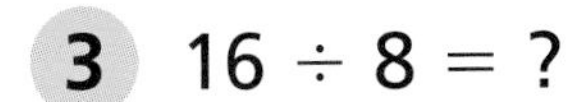

3 $16 \div 8 = ?$

There are ______ equal groups of 8. $16 \div 8 =$ ______

4 $21 \div 7 = ?$

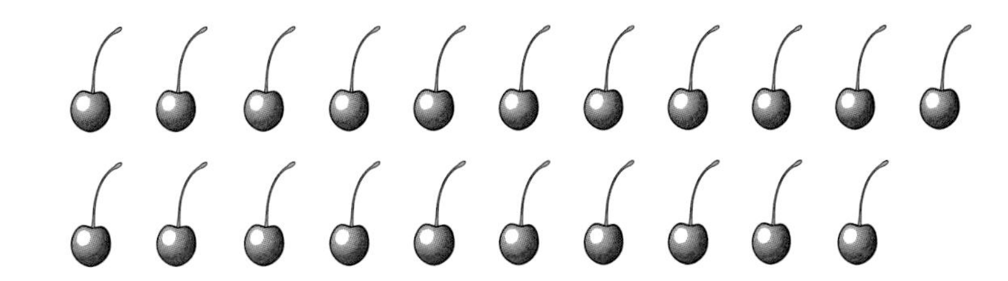

There are ______ equal groups of 7. $21 \div 7 =$ ______

5 $20 \div 5 = ?$

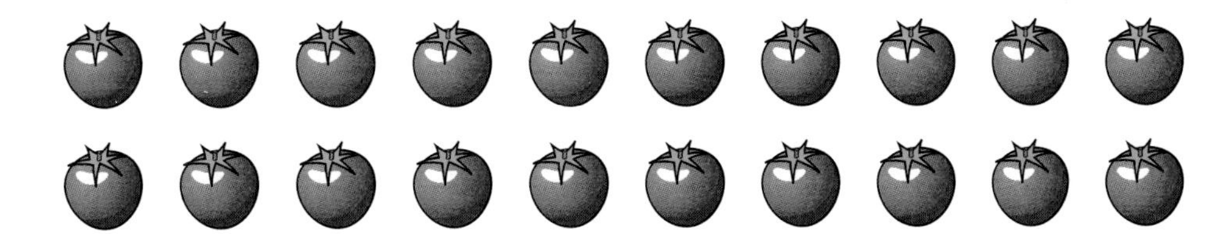

There are ______ equal groups of 5. $20 \div 5 =$ ______

6 $18 \div 2 = ?$

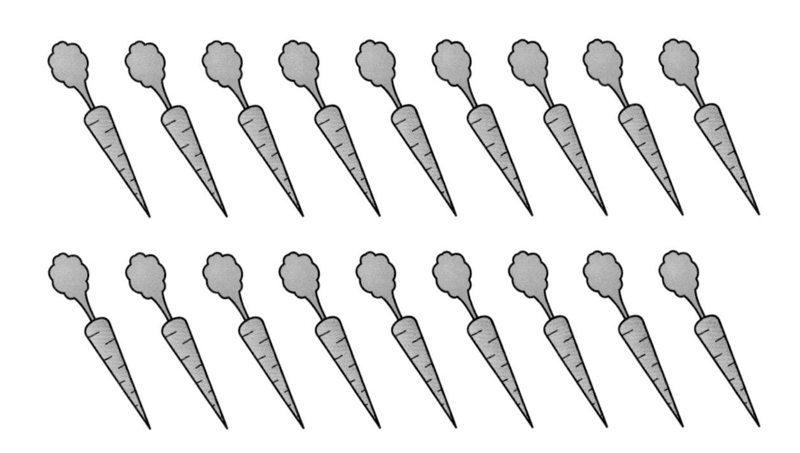

There are ______ equal groups of 2. $18 \div 2 =$ ______

Use Multiplication Facts to Divide

When you divide, you make equal groups. Whatever is left over is called the **remainder**.

▶ **Use multiplication facts to help you find the quotient. Then use the model to check your work.**

Four students share a box of 33 markers equally. How many markers does each student get?

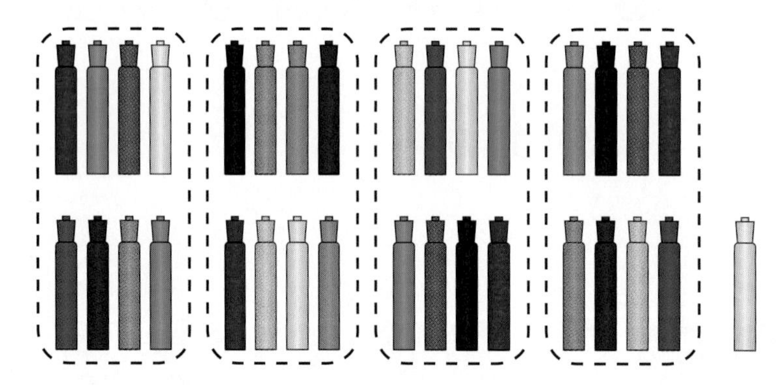

$$\begin{array}{r} 8\text{ R}1 \\ 4\overline{)33} \\ -32 \\ \hline 1 \end{array}$$

Think: What can you multiply by 4 to get close to 33 without going over?

$4 \times 8 = 32$ $4 \times 9 = 36$

- Write 8 in the quotient.
- Multiply. $4 \times 8 = 32$.
- Subtract. $33 - 32 = 1$.

Each student gets ______ markers.

There is ______ marker left over. This marker is called the **remainder**.

Use multiplication facts to help you find the quotient. Then use the model to check your work.

1. Four students share 19 tubes of paint equally. How many tubes of paint does each student get?

$4\overline{)19}$

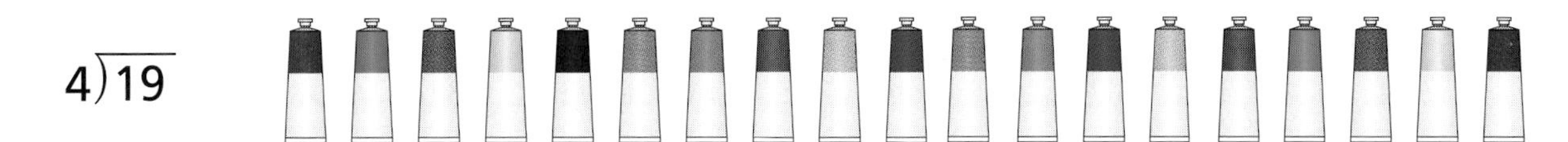

$4 \times 4 =$ ______ $4 \times 5 =$ ______

Each student gets ______ tubes of paint.

There are ______ tubes of paint left over.

Use multiplication facts to help you find each quotient.

2. $5\overline{)37}$ R
3. $9\overline{)55}$ R
4. $8\overline{)70}$ R
5. $7\overline{)44}$ R
6. $3\overline{)25}$ R
7. $6\overline{)35}$ R
8. $7\overline{)50}$ R
9. $9\overline{)39}$ R
10. $6\overline{)67}$ R
11. $5\overline{)44}$ R
12. $7\overline{)59}$ R
13. $4\overline{)39}$ R

Lesson 18 Find Quotients

When you divide, you make equal groups. Sometimes, there is a remainder.

Find the quotient. Use the model to check your work.

Three students share 38 marbles equally. How many marbles does each student get?

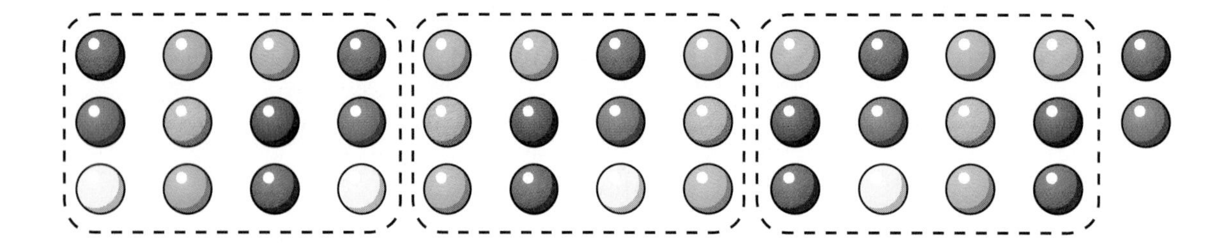

```
   12R2
3)38
 -3
  08
  -6
   2
```

- Divide 3 tens into equal groups of 3.
- Multiply. $3 \times 1 = 3$.
- Subtract. $3 - 3 = 0$.
- Bring down the 8 ones.
- Divide 8 ones into equal groups of 3.
- Multiply. $3 \times 2 = 6$.
- Subtract. $8 - 6 = 2$.

Each student gets ______ marbles.

There are ______ marbles left over. These marbles are called the **remainder**.

▶ **Find the quotient. Use the model to check your work.**

1 Three stamp collectors share 45 stamps equally. How many stamps does each person get?

$3\overline{)45}$

Each stamp collector gets ______ stamps.

There are ______ stamps left over.

▶ **Find each quotient. Three of the quotients have no remainder.**

2 $2\overline{)22}$ **3** $6\overline{)74}$ **4** $3\overline{)64}$ **5** $4\overline{)52}$

6 $3\overline{)83}$ **7** $3\overline{)46}$ **8** $7\overline{)84}$ **9** $2\overline{)39}$

▶ **Write the quotients for the problems above that had no remainder. Add to check your answers.**

10 ______ + ______ + ______ = 36

Checkup

Use repeated subtraction to find each quotient.

1. $16 \div 2 =$ ______

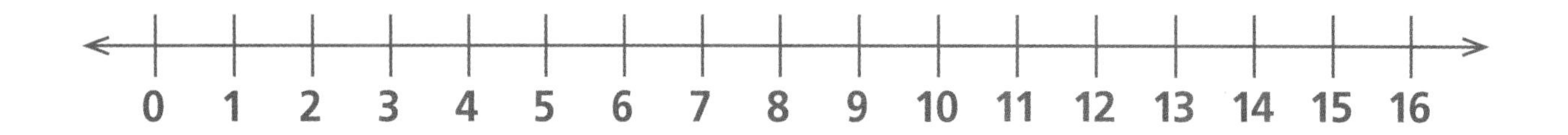

2. $20 \div 4 =$ ______

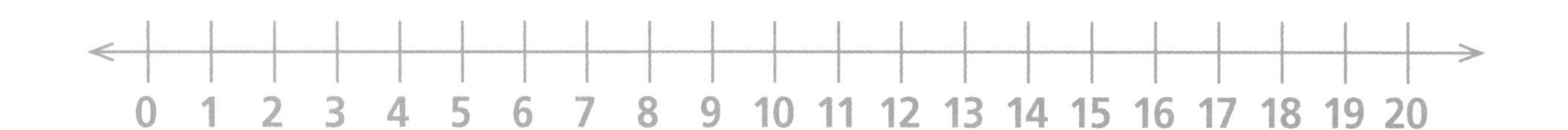

Draw circles to make equal groups. Then fill in the blanks.

3. $21 \div 3 = ?$

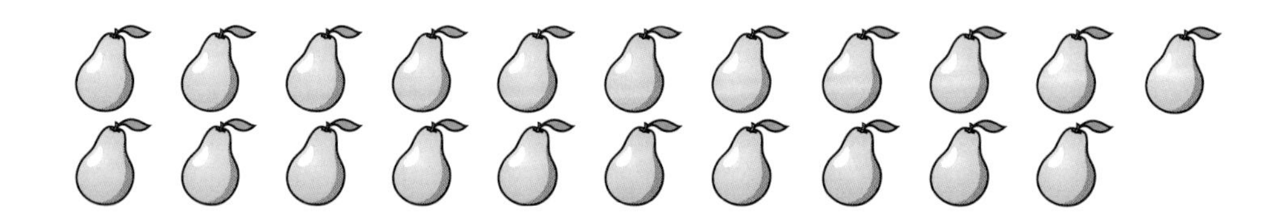

There are ______ equal groups of 3. $21 \div 3 =$ ______

4. $32 \div 8$

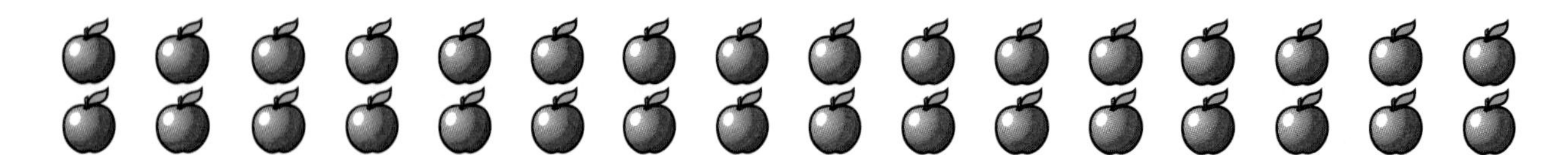

There are ______ equal groups of 8. $32 \div 8 =$ ______

Find the quotient. Use the model to check your work.

5 Seven students share 25 craft sticks equally.
How many sticks does each student get?

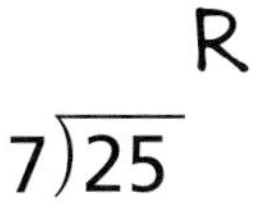

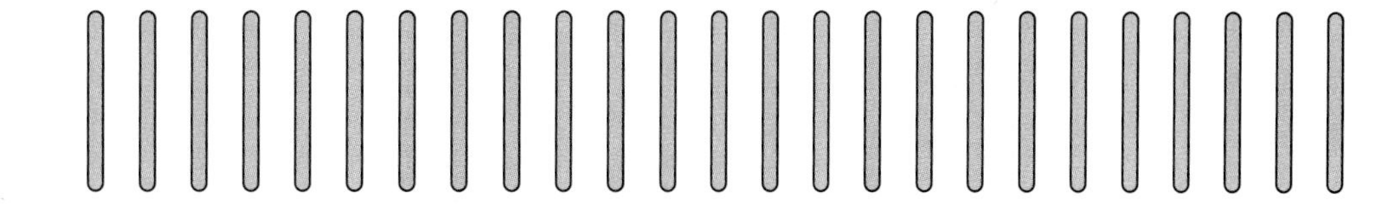

Each student gets ______ sticks. There are ______ sticks left over.

6 Two students share 21 shells equally.
How many shells does each student get?

R
2)21

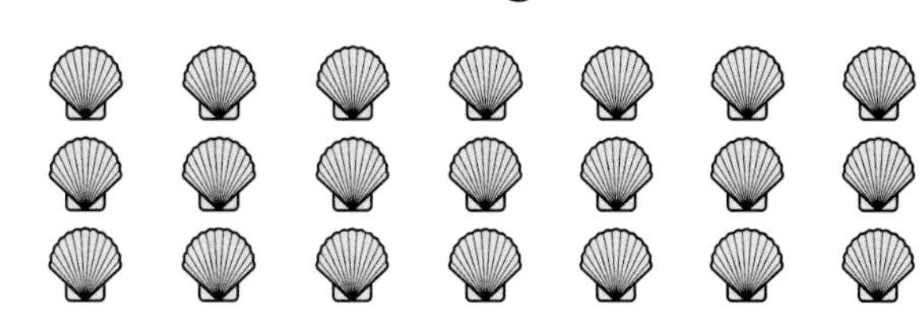

Each student gets ______ shells. There is ______ shell left over.

Use multiplication facts to help you find each quotient.

7 3)22 R

8 8)62 R

9 9)38 R

10 5)47 R

Find each quotient.

11 3)34

12 5)84

13 6)73

14 3)44

Lesson 19

Compare and Order Fractions

You can use number lines to compare fractions.

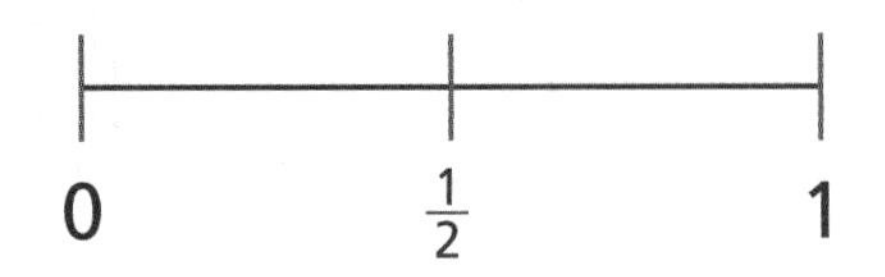

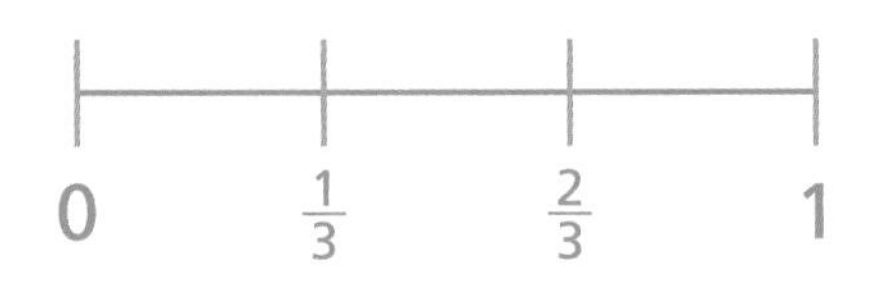

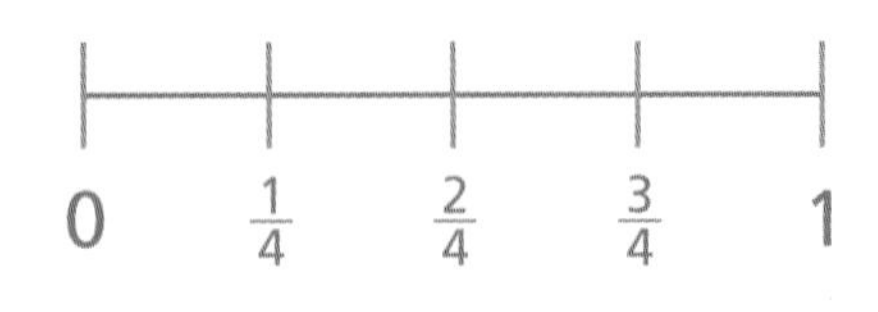

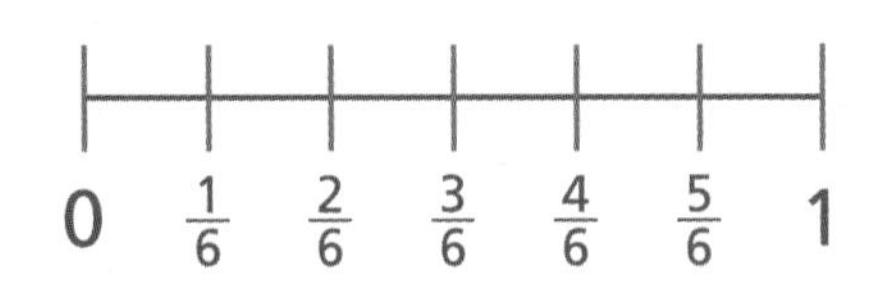

$\frac{1}{4}$ (<) $\frac{1}{2}$

$\frac{1}{4}$ is less than $\frac{1}{2}$.

$\frac{1}{2}$ (>) $\frac{1}{3}$

$\frac{1}{2}$ is greater than $\frac{1}{3}$.

$\frac{1}{3}$ (=) $\frac{2}{6}$

$\frac{1}{3}$ equals $\frac{2}{6}$.

▶ **Compare the fractions. Write $>$, $<$, or $=$ in the circle.**

$\frac{1}{4}$ ◯ $\frac{1}{3}$ $\frac{1}{2}$ ◯ $\frac{3}{6}$ $\frac{2}{4}$ ◯ $\frac{1}{2}$

You can also use number lines to order fractions.

▶ **Mark each fraction on the number lines above.**
Then write the fractions in order from least to greatest.

Fractions: $\frac{2}{3}$, $\frac{1}{6}$, $\frac{1}{2}$, $\frac{3}{4}$

The fractions from
least to greatest are $\frac{1}{6}$, ______, ______, $\frac{3}{4}$.

Compare the fractions. Write >, <, or = in the circle.

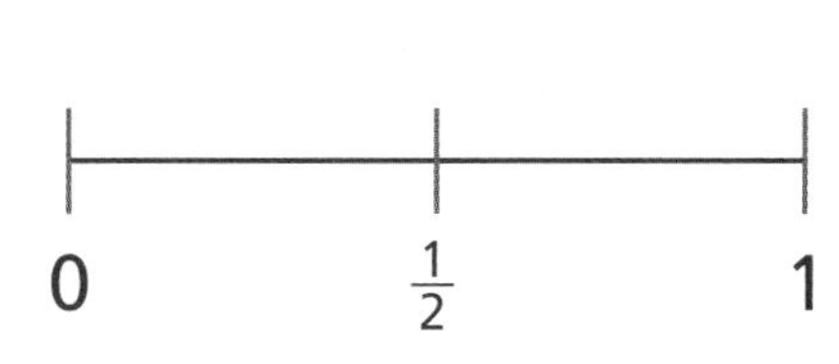

0 $\frac{1}{5}$ $\frac{2}{5}$ $\frac{3}{5}$ $\frac{4}{5}$ 1

0 $\frac{1}{8}$ $\frac{2}{8}$ $\frac{3}{8}$ $\frac{4}{8}$ $\frac{5}{8}$ $\frac{6}{8}$ $\frac{7}{8}$ 1

0 $\frac{1}{10}$ $\frac{2}{10}$ $\frac{3}{10}$ $\frac{4}{10}$ $\frac{5}{10}$ $\frac{6}{10}$ $\frac{7}{10}$ $\frac{8}{10}$ $\frac{9}{10}$ 1

1 $\frac{5}{8}$ ◯ $\frac{2}{5}$

2 $\frac{1}{2}$ ◯ $\frac{7}{8}$

3 $\frac{2}{5}$ ◯ $\frac{4}{5}$

4 $\frac{7}{10}$ ◯ $\frac{1}{5}$

5 $\frac{1}{2}$ ◯ $\frac{5}{10}$

6 $\frac{6}{8}$ ◯ $\frac{3}{5}$

Mark each fraction on the number lines above. Then write the fractions in order from least to greatest.

7 $\frac{5}{8}$, $\frac{1}{10}$, $\frac{3}{8}$, $\frac{2}{5}$ ________, ________, ________, ________

8 $\frac{1}{2}$, $\frac{3}{10}$, $\frac{1}{5}$, $\frac{5}{8}$ ________, ________, ________, ________

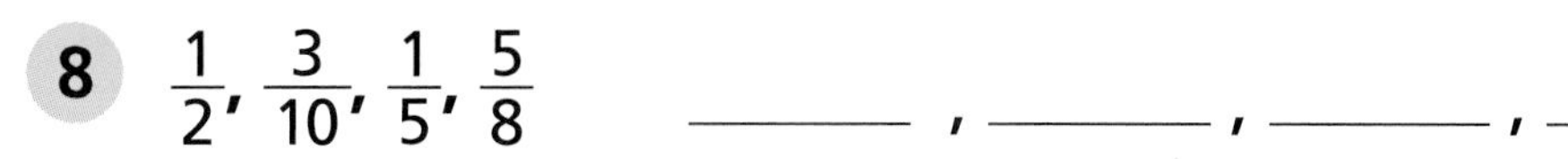

Lesson 20

Equivalent Fractions

A fraction has two parts. The **denominator** tells how many parts there are in all. The **numerator** tells how many parts you are talking about.

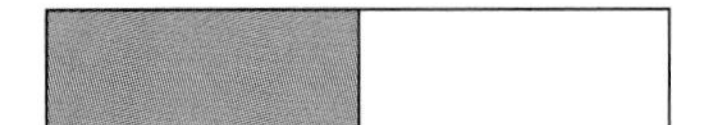

1 out of 2 parts is shaded.

$\frac{1}{3}$ and $\frac{2}{6}$ are the same amount. They are **equivalent fractions**.

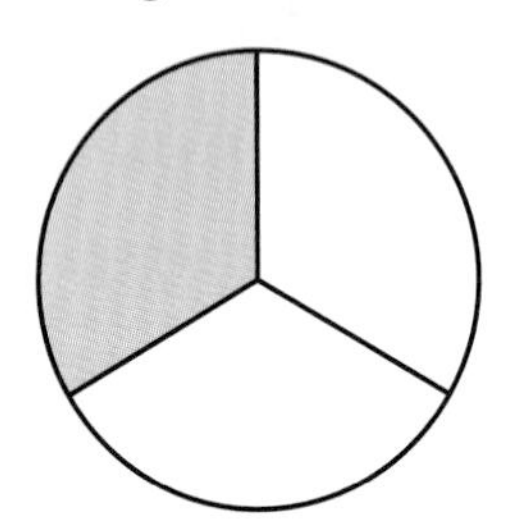 = 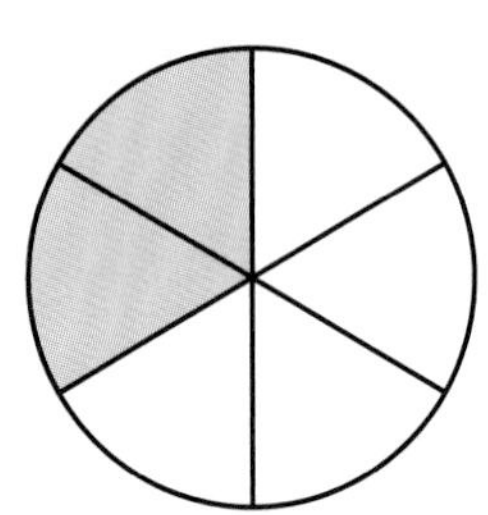

$\frac{1}{3} = \frac{\_\_}{6}$

▶ **Use the models to write equivalent fractions.**

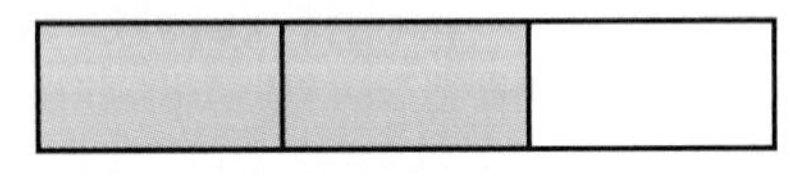

$\frac{2}{3} = \frac{\_\_}{6}$

▶ **Use the models to write equivalent fractions.**

1

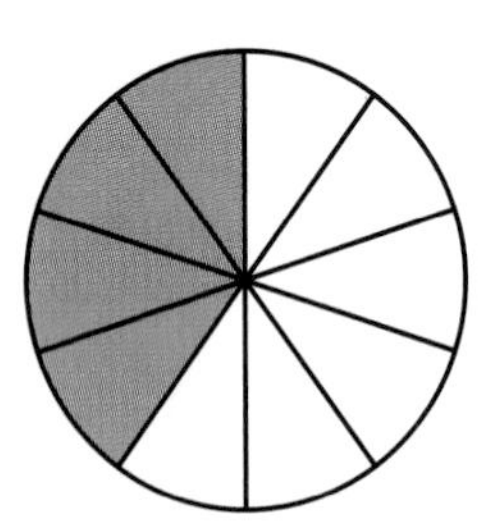 = 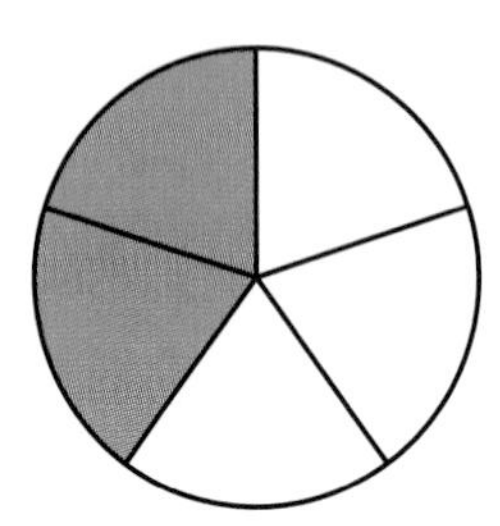

$\frac{\_\_}{10} = \frac{\_\_}{5}$

2

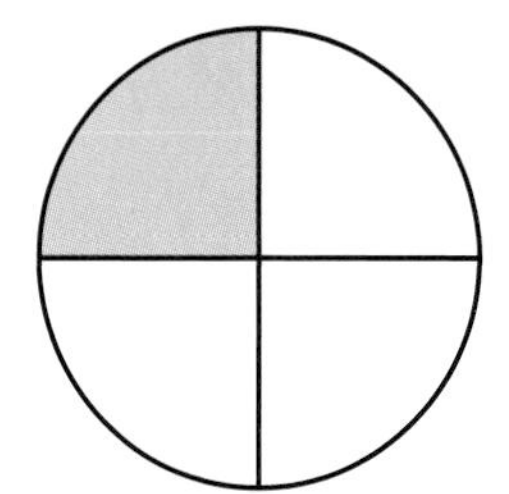

=

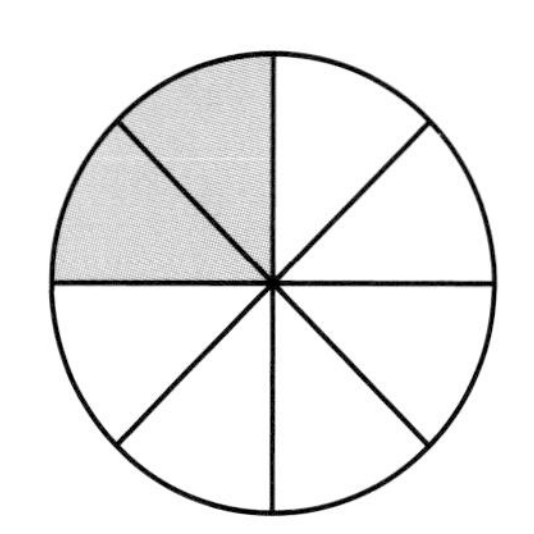

$\frac{\_\_\_}{4} = \frac{\_\_\_}{8}$

3

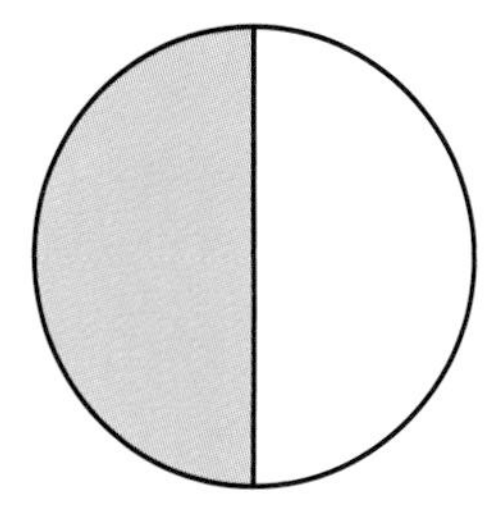

=

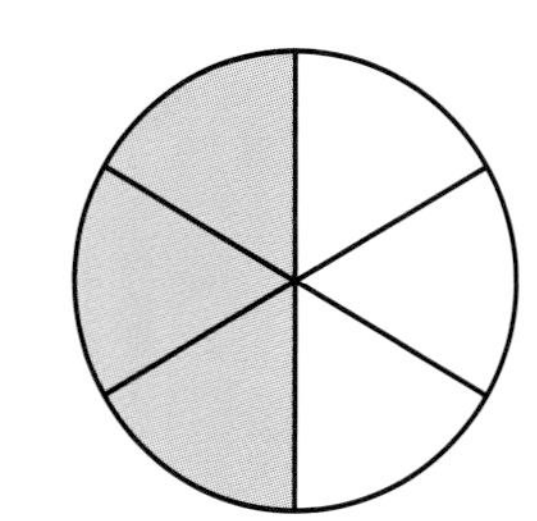

$\frac{\_\_\_}{2} = \frac{\_\_\_}{6}$

4

$\frac{\_\_\_}{3} = \frac{\_\_\_}{6}$

5

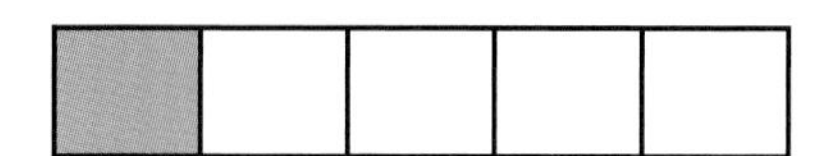

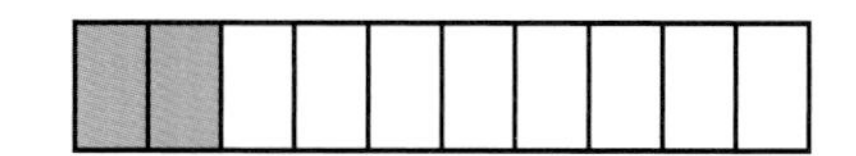

$\frac{\_\_\_}{5} = \frac{\_\_\_}{10}$

Lesson

Add and Subtract Fractions

Use the model to add the fractions.

$\frac{1}{5} + \frac{2}{5} = \frac{3}{5}$

$\frac{1}{10} + \frac{4}{10} = \frac{\_\_\_}{10}$

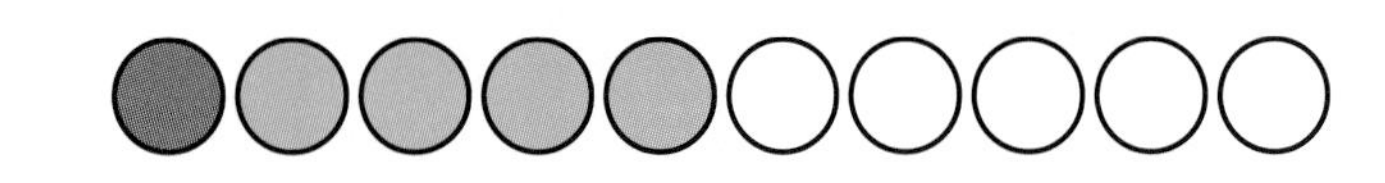

Use the model to add the fractions.

1. $\frac{3}{8} + \frac{2}{8} = \frac{\_\_\_}{8}$

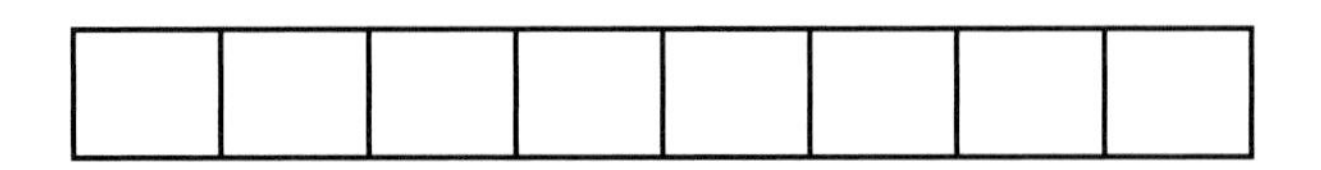

2. $\frac{2}{3} + \frac{1}{3} = $ ______

3. $\frac{3}{5} + \frac{1}{5} = $ ______

4. $\frac{2}{6} + \frac{2}{6} = $ ______

Use the model to subtract the fractions.

$\frac{5}{8} - \frac{2}{8} = \frac{3}{8}$

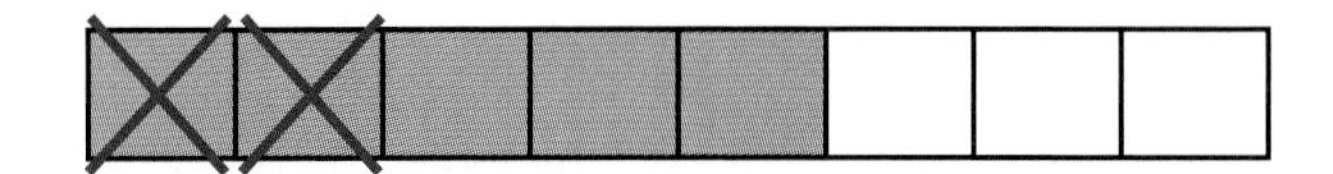

$\frac{6}{6} - \frac{2}{6} = \frac{\_\_\_}{6}$

Use the model to subtract the fractions.

5 $\frac{3}{4} - \frac{1}{4} = \frac{\_\_\_}{4}$

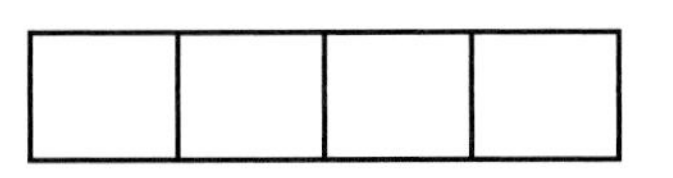

6 $\frac{6}{7} - \frac{2}{7} = \frac{\_\_\_}{7}$

7 $\frac{3}{3} - \frac{1}{3} = \frac{\_\_\_}{3}$

8 $\frac{7}{10} - \frac{4}{10} = \frac{\_\_\_}{10}$

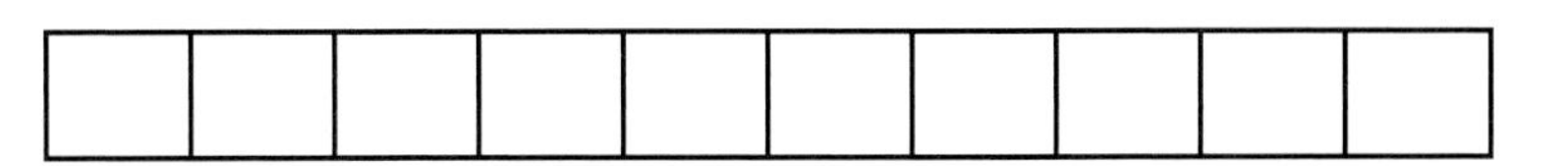

Lesson 22 Add and Subtract Money

A money amount can be written as a **decimal**.

1¢	5¢	10¢	25¢	50¢	100¢
\$0.01	\$0.05	\$0.10	\$0.25	\$0.50	\$1.00

↖ **decimal point**

You can add and subtract money. Always line up the decimal points first. Then add or subtract as you would whole numbers.

Find the sum. \$1.25 + \$0.25

- Add the ones, and regroup.

$$\begin{array}{r} {}^{1} \\ \$1.25 \\ +\$0.25 \\ \hline 0 \end{array}$$

- Add all the tens.

$$\begin{array}{r} {}^{1} \\ \$1.25 \\ +\$0.25 \\ \hline .50 \end{array}$$

- Add the hundreds.

$$\begin{array}{r} {}^{1} \\ \$1.25 \\ +\$0.25 \\ \hline \$1.50 \end{array}$$

Find each sum.

1
$$\begin{array}{r} \$3.25 \\ +\$1.19 \\ \hline \end{array}$$

2
$$\begin{array}{r} \$2.63 \\ +\$1.47 \\ \hline \end{array}$$

3
$$\begin{array}{r} \$3.21 \\ +\$2.89 \\ \hline \end{array}$$

Find the difference. $2.50 − $0.75

- Regroup 1 ten as 10 ones. Then subtract the ones.

$$\begin{array}{r} \scriptstyle 4\,10 \\ \$\,2.\not{5}\not{0} \\ -\ \$\,0.7\,5 \\ \hline 5 \end{array}$$

- Regroup 1 hundred as 10 tens. Then subtract the tens.

$$\begin{array}{r} \scriptstyle 14 \\ \scriptstyle 1\ \not{4}10 \\ \$\not{2}.\not{5}\not{0} \\ -\ \$\,0.7\,5 \\ \hline .75 \end{array}$$

- Subtract the hundreds.

$$\begin{array}{r} \scriptstyle 14 \\ \scriptstyle 1\ \not{4}10 \\ \$\not{2}.\not{5}\not{0} \\ -\ \$\,0.7\,5 \\ \hline \$1.75 \end{array}$$

Find each difference.

4 $\begin{array}{r} \$0.59 \\ -\$0.24 \\ \hline \end{array}$

5 $\begin{array}{r} \$2.19 \\ -\$1.70 \\ \hline \end{array}$

6 $\begin{array}{r} \$1.45 \\ -\$0.77 \\ \hline \end{array}$

Add or subtract to solve each problem.

7 $\begin{array}{r} \$3.60 \\ +\$1.98 \\ \hline \end{array}$

8 $\begin{array}{r} \$6.07 \\ +\$2.99 \\ \hline \end{array}$

9 $\begin{array}{r} \$2.45 \\ -\$1.50 \\ \hline \end{array}$

10 $\begin{array}{r} \$5.00 \\ -\$2.85 \\ \hline \end{array}$

Checkup

Compare the fractions. Write >, <, or = in the circle.

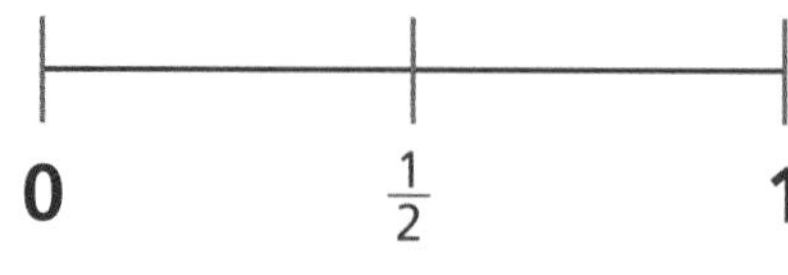

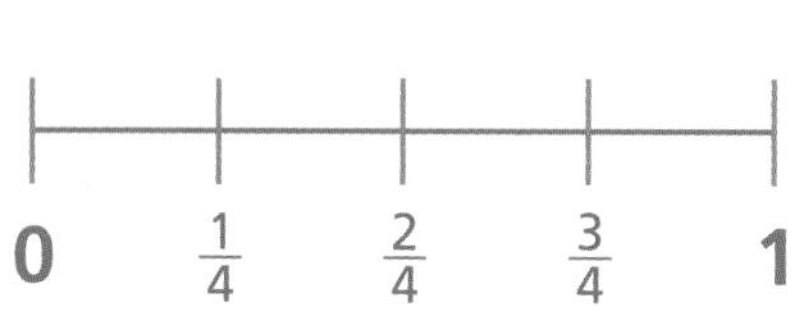

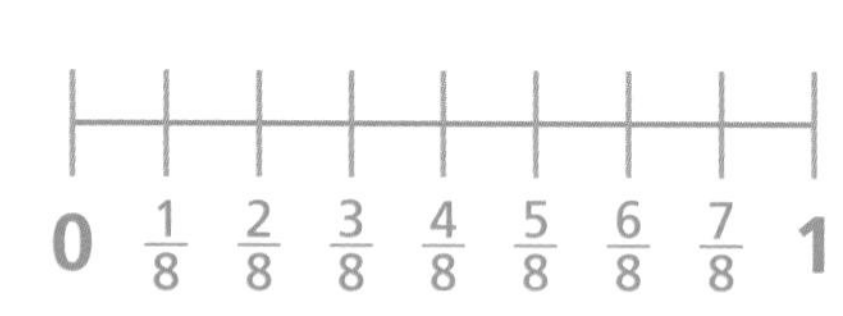

1. $\frac{1}{4}$ ○ $\frac{7}{8}$

2. $\frac{5}{8}$ ○ $\frac{1}{2}$

3. $\frac{1}{2}$ ○ $\frac{2}{4}$

Use the number lines above to order the fractions from least to greatest.

4. $\frac{1}{2}$, $\frac{3}{8}$, $\frac{3}{4}$ ______, ______, ______

Use the model to write equivalent fractions.

5.

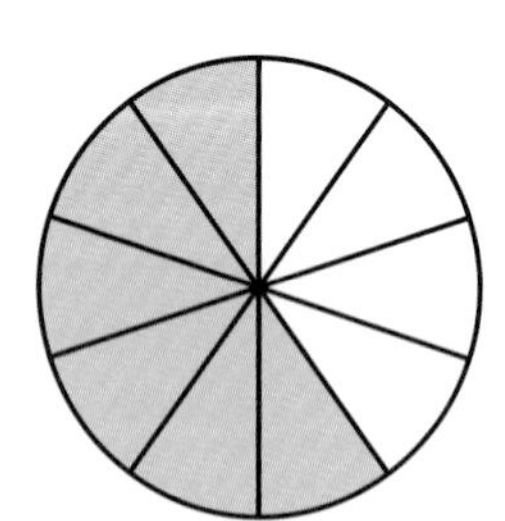 = 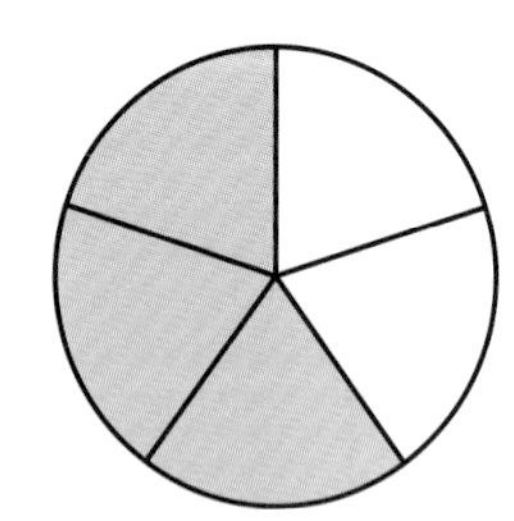

$\frac{\_\_\_}{10} = \frac{\_\_\_}{5}$

6.

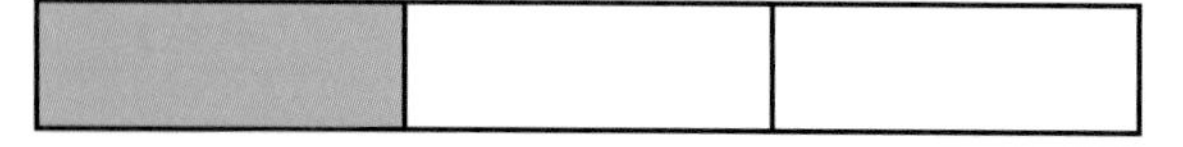

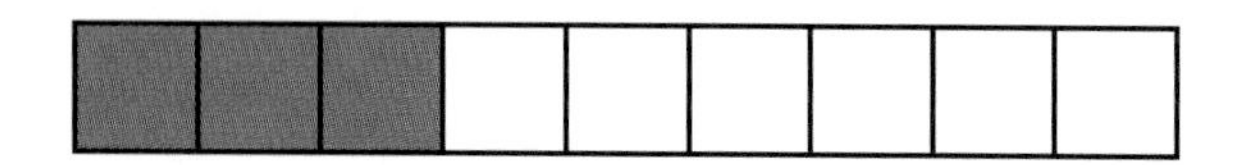

$\frac{\_\_\_}{3} = \frac{3}{9}$

Use the model to add the fractions.

7. $\frac{2}{4} + \frac{1}{4} = \frac{\quad}{4}$

8. $\frac{1}{5} + \frac{3}{5} = \frac{\quad}{5}$

Use the model to subtract the fractions.

9. $\frac{4}{5} - \frac{2}{5} = \frac{\quad}{5}$

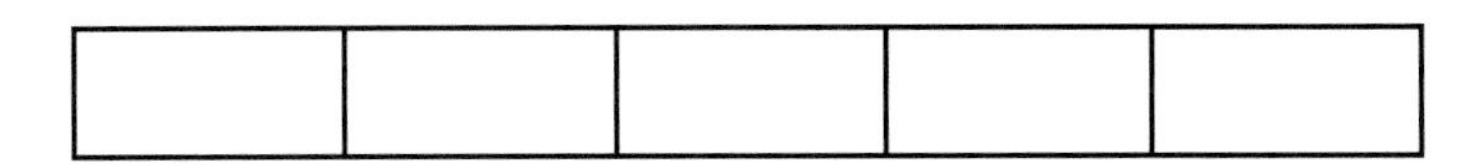

10. $\frac{7}{9} - \frac{5}{9} = \frac{\quad}{9}$

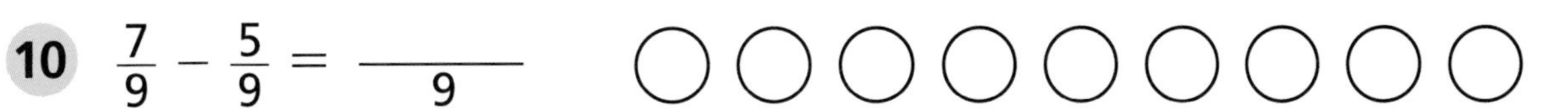

Find each sum.

11. $3.55 + $2.99 = ____

12. $2.98 + $1.05 = ____

13. $5.99 + $3.99 = ____

Find each difference.

14. $0.91 − $0.56 = ____

15. $2.00 − $0.99 = ____

16. $1.50 − $0.75 = ____

Number Patterns

Write what comes next in the pattern. Then write the rule.

49, 42, 35, ______, ______, ______

- Do the numbers increase or decrease? ______________
- Does each pair of numbers change by the same amount? ______________
- What is the rule for this pattern? ______________

Complete the table. Then write the rule.

Number of triangles	1	2	3	4	5	6	7
Number of sides	3	6	9	12	15	18	21

The rule is ______________.

Find the pattern. Then write the rule.

A. Rule: ______________

1	3
2	4
3	5
4	6

B. Rule: ______________

3	6
4	8
5	10
6	12

Write the numbers that come next in the pattern. Then write the rule.

1 1, 2, 4, 6, ______, ______, ______

The rule is ______________________.

2 60, 50, ______, ______, 20, 10

The rule is ______________________.

Complete the table. Then write the rule.

3 At a fair, visitors can get 5 tickets for one dollar.

Number of dollars	1	2	3	4	5	6	7	8
Number of tickets	5	10	15	____	____	____	____	____

The rule is ______________________.

Find the pattern. Then write the rule.

4

Rule: ______________________	
1	3
2	6
3	9
4	12

5

Rule: ______________________	
20	16
40	36
60	56
80	76

Related Addition and Subtraction Facts

A fact family is a set of related facts.

Use these numbers to write a fact family: 2, 3, 5.

$2 + 3 = 5$ $3 + 2 = 5$ $5 - 2 = 3$ $5 - 3 = 2$

Use related facts to find the missing number.

How many marbles are hidden?

+ ? = 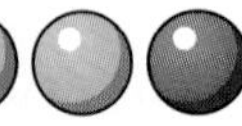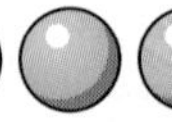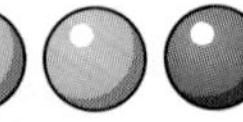

$4 +$? $= 7$

The opposite of addition is subtraction. Use a related subtraction fact to find the missing number.

$7 - 4 =$? The missing number is ______.

Use each set of numbers to write a fact family.

1 15, 2, 17 ____________, ____________

____________, ____________

2 27, 12, 15 ____________, ____________

____________, ____________

3 20, 5, 15 ____________, ____________

____________, ____________

Use related facts to find the missing numbers.

4  + ? =

3 + ? = 11 Related fact: ____________

The missing number is ______.

5 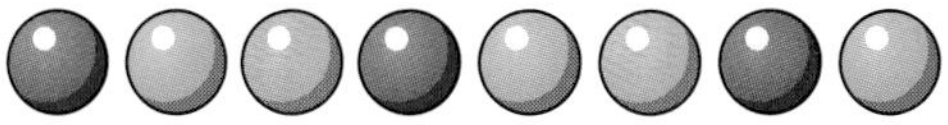 − ? =

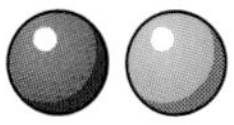

8 − ? = 2 Related fact: ____________

The missing number is ______.

6 6 + ? = 10 Related fact: ____________

The missing number is ______.

7 20 − ? = 13 Related fact: ____________

The missing number is ______.

8 ? + 14 = 19 Related fact: ____________

The missing number is ______.

9 ? − 8 = 12 Related fact: ____________

The missing number is ______.

10 11 + ? = 22 Related fact: ____________

The missing number is ______.

Related Multiplication and Division Facts

A **fact family** is a set of related facts.

Use these numbers to write a fact family: 4, 5, 20.

$4 \times 5 = 20$ $5 \times 4 = 20$ $20 \div 4 = 5$ $20 \div 5 = 4$

Use related facts to find the missing number.

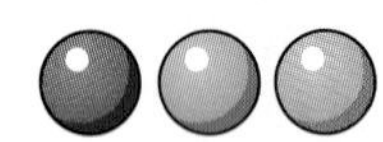 

How many marbles are in each group to make 6 in all?

$2 \times ? = 6$

The opposite of multiplication is division. Use a related division fact to find the missing number.

$6 \div 2 = ?$ The missing number is ______.

Use each set of numbers to write a fact family.

1 30, 6, 5 ______________, ______________

______________, ______________

2 7, 28, 4 ______________, ______________

______________, ______________

3 9, 5, 45 ______________, ______________

______________, ______________

Use related facts to find the missing numbers.

 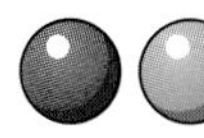 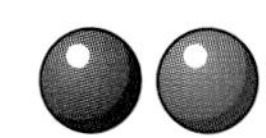 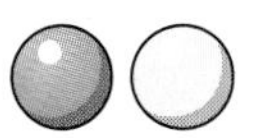 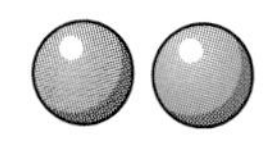

$2 \times ? = 10$ Related fact: ________

The missing number is ____.

5   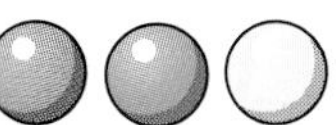 

$10 \div ? = 5$ Related fact: ________

The missing number is ____.

6 $7 \times ? = 21$ Related fact: ________

The missing number is ____.

7 $14 \div ? = 2$ Related fact: ________

The missing number is ____.

8 $? \div 5 = 2$ Related fact: ________

The missing number is ____.

9 $? \times 3 = 27$ Related fact: ________

The missing number is ____.

10 $7 \times ? = 49$ Related fact: ________

The missing number is ____.

Lesson 26

Solve Number Sentences

In an **open sentence**, one or more numbers are missing.

Replace each symbol with its number. Then solve the number sentence.

⬟ = 3 ▲ = 5

⬟ + ▲ = 3 + 5 = 8

Replace each symbol with its number. Then solve the number sentence.

▲ = 4 ■ = 5 ⬟ = 6 ⏢ = 7

1. ■ + ⏢ = ? ______ + ______ = ______

2. ⬟ − ▲ = ? ______ − ______ = ______

3. 42 ÷ ⬟ = ⏢ ______ ÷ ______ = ______

4. ⏢ − 4 = ? ______ − 4 = ______

5 $9 \times$ ■ $= ?$ $9 \times$ ______ $=$ ______

6 ⬟ $\div 2 = ?$ ______ $\div 2 =$ ______

7 $11 +$ ⏢ $= ?$ $11 +$ ______ $=$ ______

8 $17 -$ ■ $= ?$ $17 -$ ______ $=$ ______

9 ▲ $+$ ⏢ $= ?$ ______ $+$ ______ $=$ ______

10 ⬟ $+ 5 = ?$ ______ $+ 5 =$ ______

11 ▲ $\times$ ■ $= ?$ ______ $\times$ ______ $=$ ______

12 $100 \div$ ■ $= ?$ $100 \div$ ______ $=$ ______

13 ▲ $+$ ⏢ $+ 1 = ?$ ______ $+$ ______ $+ 1 =$ ______

14 ⏢ $- 2 = 1 +$ ▲ ______ $- 2 = 1 +$ ______

15 ⬟ $\times 2 =$ ■ $+$ ⏢ ______ $\times 2 =$ ______ $+$ ______

Checkup

Write what comes next in the pattern. Then write the rule.

1. 10, 15, ______ , ______ , 30

   The rule is ____________________.

Complete the table. Then write the rule.

2. A toy truck has 6 wheels.

Number of toy trucks	1	2	3	4	5	6
Number of wheels	6	12	____	____	____	____

The rule is ____________________.

Find the pattern. Then write the rule.

3. Rule: ____________________

40	35
30	25
20	15
10	5

4. Rule: ____________________

3	6
6	12
9	18
12	24

Use the set of numbers to write an addition and subtraction fact family.

5. 13, 7, 20 ____________________ , ____________________

   ____________________ , ____________________

Use a related fact to find the missing number.

6 $18 + ? = 25$ Related fact: ____________________

The missing number is ______.

Use the set of numbers to write a multiplication and division fact family.

7 8, 2, 16 ____________________ , ____________________

____________________ , ____________________

Use a related fact to find the missing number.

8 $? \times 7 = 42$ Related fact: ____________________

The missing number is ______.

Replace each symbol with its number. Then solve the number sentence.

▲ = 5 ■ = 8 ⏢ = 9

9 ■ + ▲ = ? ______ + ______ = ______

10 ■ + 12 = ? ______ + 12 = ______

11 ⏢ × 8 = ? ______ × 8 = ______

12 ▲ + ■ + ⏢ = ? ______ + ______ + ______ = ______

13 ⏢ + 2 + ▲ = ? ______ + 2 + ______ = ______

Lesson 27

Lines

A line is a straight path that goes on forever in opposite directions.

A line segment is part of a line. It has an endpoint on each end.

Parallel lines never meet. They are always the same distance from each other.

Perpendicular lines cross.

An angle forms where two lines or line segments meet or cross.
A right angle measures 90°.
Perpendicular lines form right angles.

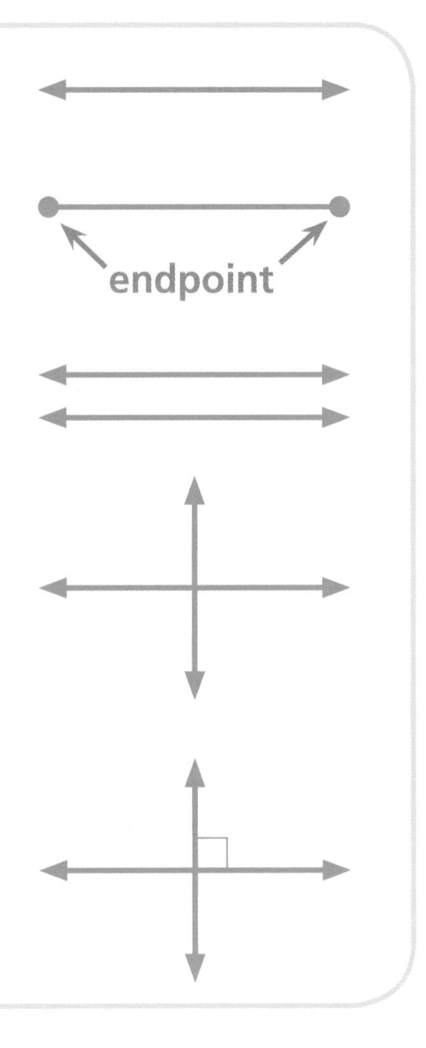

Draw lines to match each word to its example.

1 line segment

2 line

3 perpendicular lines

4 parallel lines

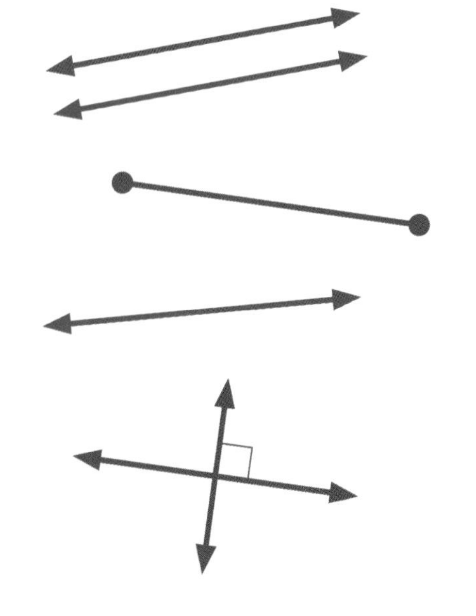

Write the letter of the figure that the words describe.

5 ______ This figure has two endpoints.

A. 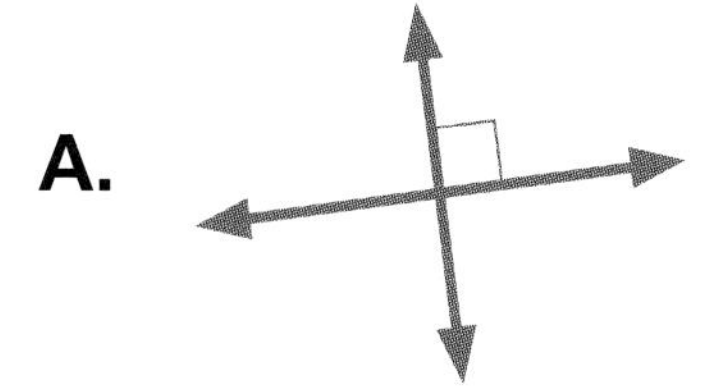

6 ______ Where these lines meet, they form right angles.

B. 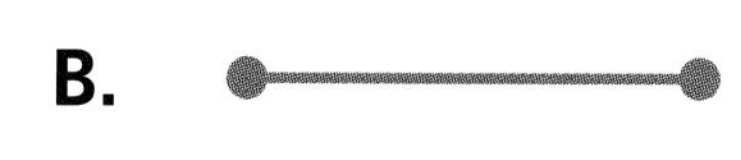

7 ______ This figure goes on forever in two directions.

C. 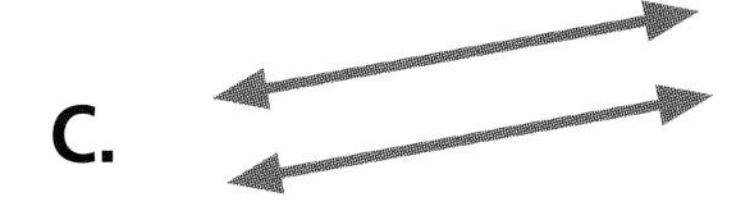

8 ______ These two lines always remain the same distance apart.

D. 

Draw an example of each figure.

9	line	
10	line segment	
11	parallel lines	
12	perpendicular lines	

Polygons

A **polygon** is a closed figure made of 3 or more line segments.

A **triangle** is a polygon. It has 3 sides, or line segments, and 3 angles.

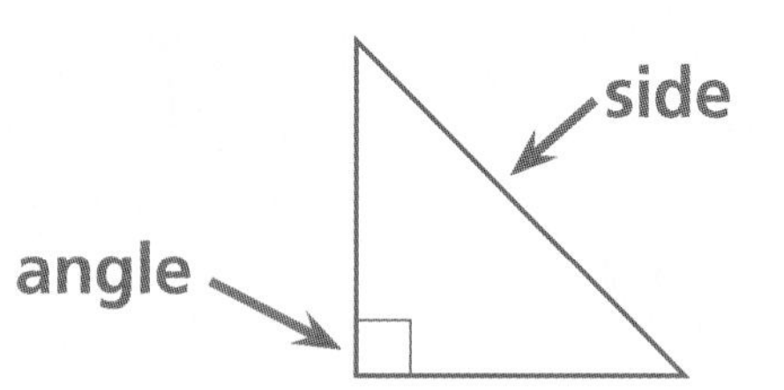

Here are some more polygons.

quadrilateral	pentagon	hexagon	octagon
Examples:	Examples:	Examples:	Examples:
4 sides	5 sides	6 sides	8 sides
4 angles	5 angles	6 angles	8 angles

▶ **Draw lines to match each polygon to its example.**

1 triangle

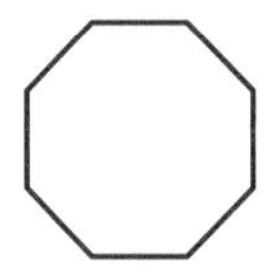

2 quadrilateral

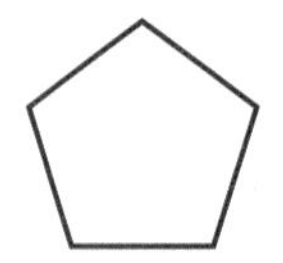

3 pentagon

4 hexagon

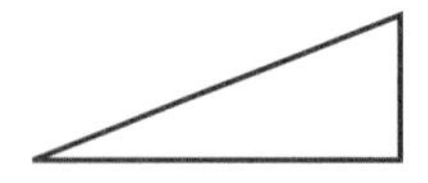

5 octagon

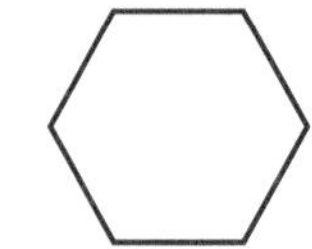

Write the letter of the polygon that the words describe.

6	______ 8 sides and 8 angles	A.
7	______ 5 sides and 5 angles	B.
8	______ 4 sides and 4 angles	C.

Fill in the chart.

	Polygon	Name	Number of Sides	Number of Angles
9		rectangle		
10				4
11			3	
12		pentagon		
13		hexagon		
14			8	
15		decagon		

Lesson 29

Quadrilaterals

A **quadrilateral** is a polygon with 4 sides and 4 angles.
Here are some examples.

rectangle	square	rhombus	trapezoid
Examples:	Examples:	Examples:	Examples:
4 sides	4 equal sides	4 equal sides	4 sides
2 pairs of parallel sides	2 pairs of parallel sides	2 pairs of parallel sides	1 pair of parallel sides
4 right angles	4 right angles	0 right angles	0, 1, or 2 right angles

Circle each example of a quadrilateral.

1

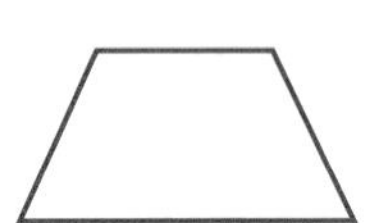 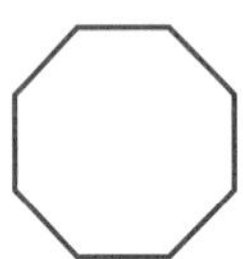  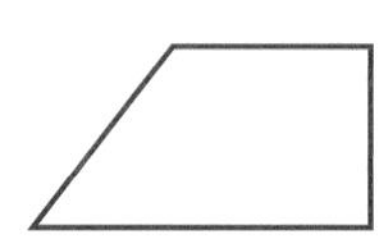

Draw lines to match each quadrilateral to its example.

2 rectangle

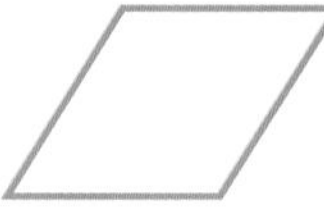

3 trapezoid

4 square

5 rhombus

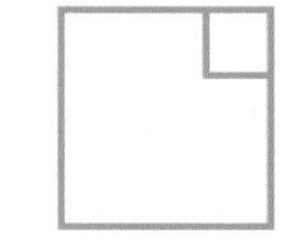

Decide if each sentence is TRUE or FALSE. Circle your answer.

6. A triangle is a quadrilateral. TRUE FALSE

7. A square has 2 pairs of parallel sides of equal length. TRUE FALSE

8. A quadrilateral always has 2 pairs of parallel sides. TRUE FALSE

9. A rhombus has 4 right angles. TRUE FALSE

Draw an example of each quadrilateral.

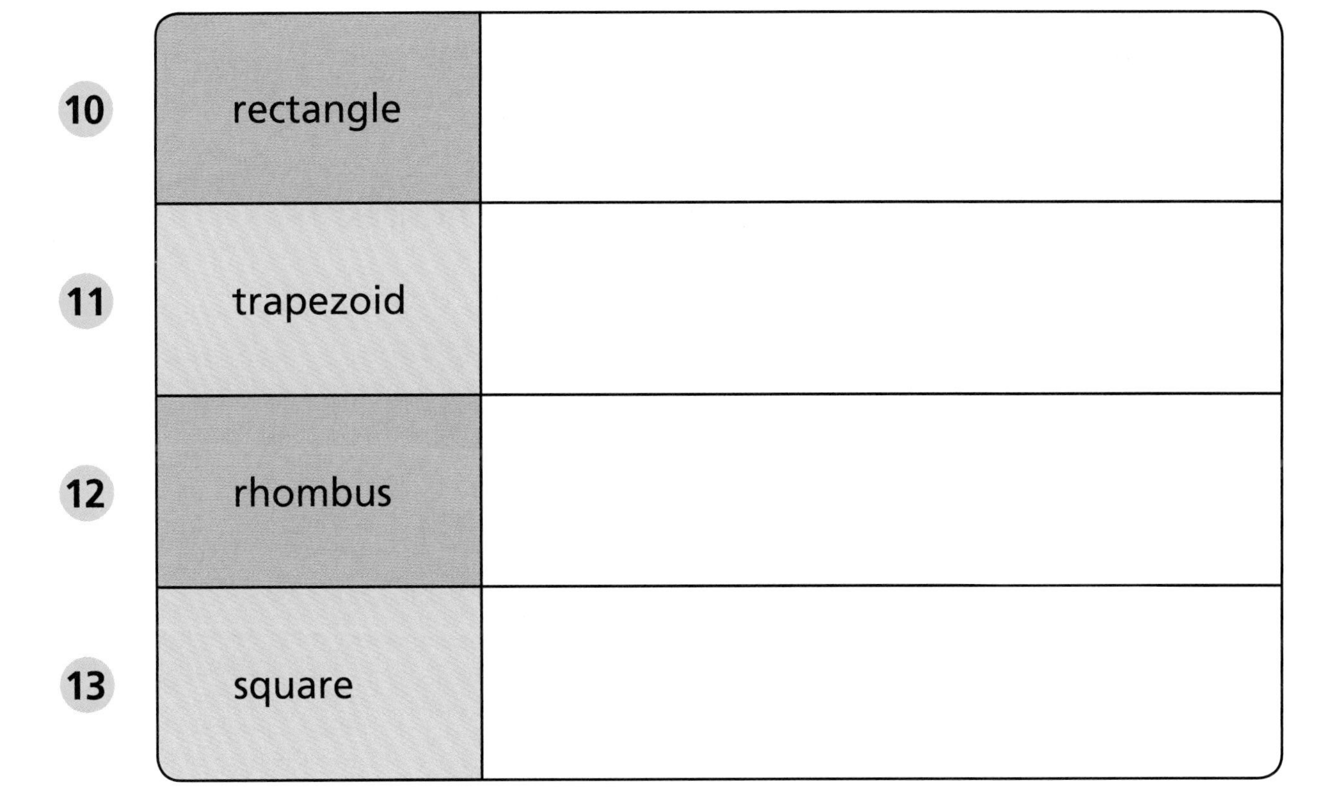

10	rectangle	
11	trapezoid	
12	rhombus	
13	square	

Circle ways these quadrilaterals are the same.

14

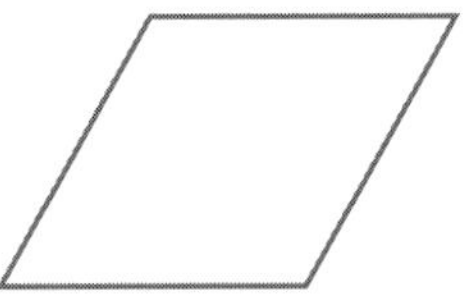

4 equal sides 4 angles 4 right angles

Lesson 30

Solid Figures

Solid figures are 3-dimensional because they have length, width, and height.

A **pyramid** is an example.

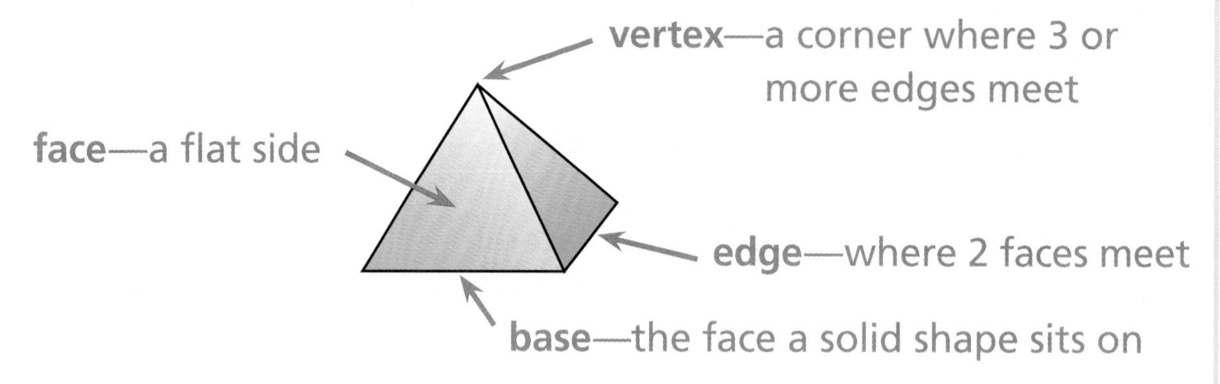

A pyramid has 5 faces, 8 edges, and 5 vertices.

Spheres, cones, and cylinders are also solid figures.

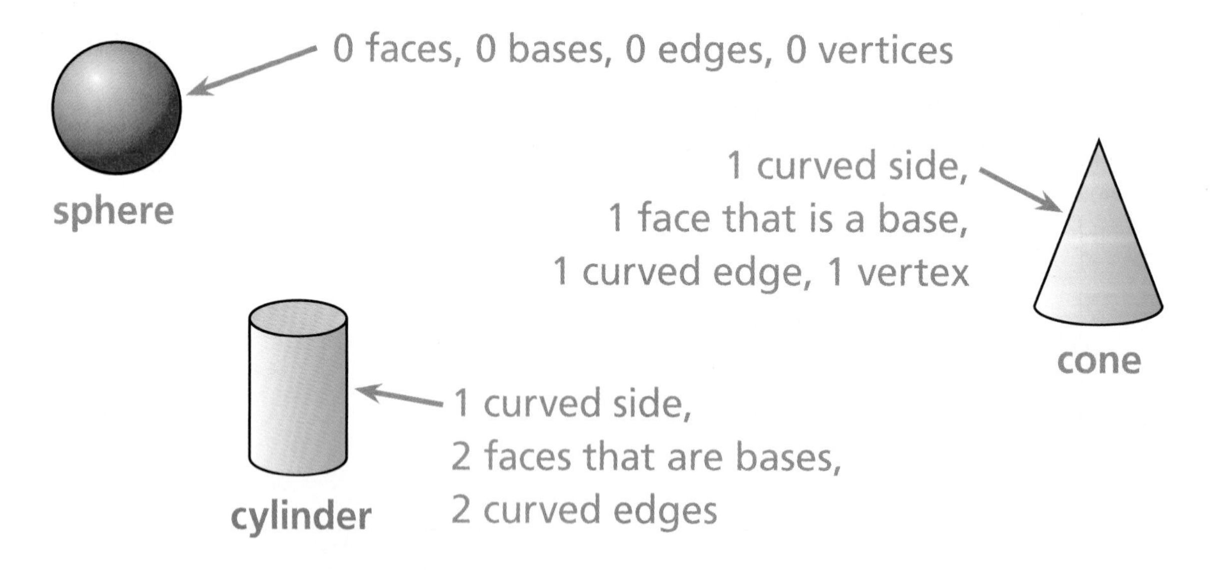

Circle the solid figures.

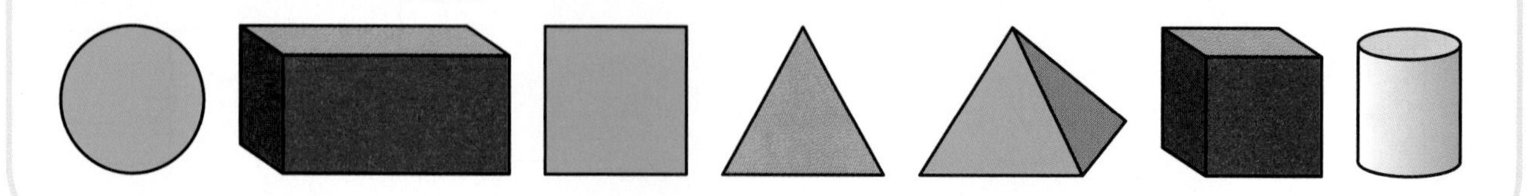

▶ **Write the letter of the example that matches each name.**

1	______ sphere	A.	
2	______ pyramid	B.	SOUP
3	______ cylinder	C.	
4	______ cone	D.	

▶ **Fill in the blanks.**

	Solid Figure	Name	Number of Curved Sides	Number of Faces	Number of Edges	Number of Vertices
5		sphere				
6		pyramid	0			
7					1	1
8		cylinder	1			

Prisms

A **prism** is a solid figure that gets its name from the shape of one or more of its faces.

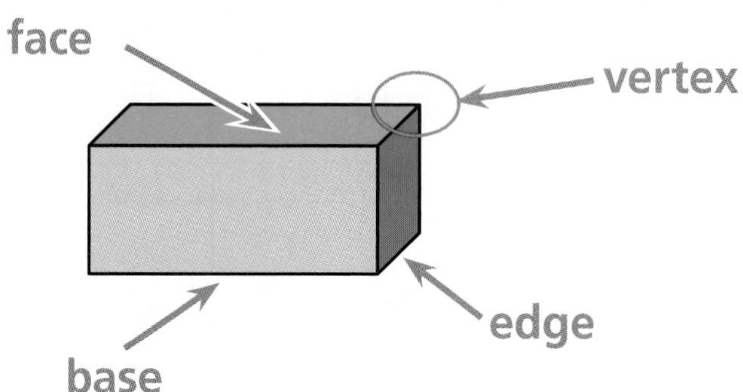

All of the faces of this solid figure are rectangles. So, this figure is called a **rectangular prism**.

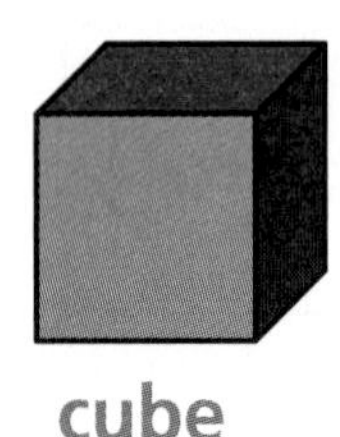
cube

This solid figure is a type of rectangular prism.
Each of its faces is a square.
Each square is the same size.
So, this figure is called a **cube**.

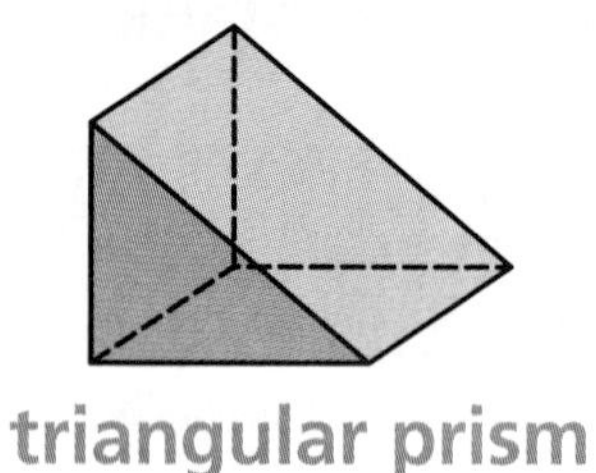
triangular prism

This solid figure has two triangular faces.
So, this figure is called a **triangular prism**.

Draw lines to match each word to its example.

1 rectangular prism

2 cube

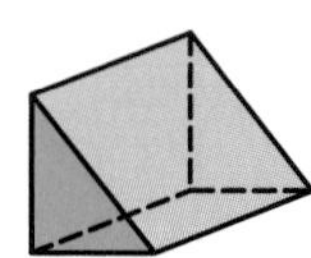

3 triangular prism

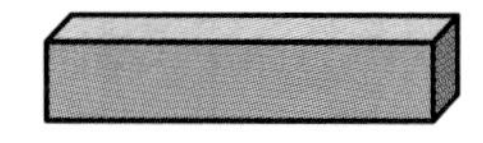

Write the letter of the example that matches each name.

4	_____ rectangular prism	A.	
5	_____ triangular prism	B.	
6	_____ cube	C.	

Use these words to label the parts of the prism.

7 face base edge vertex

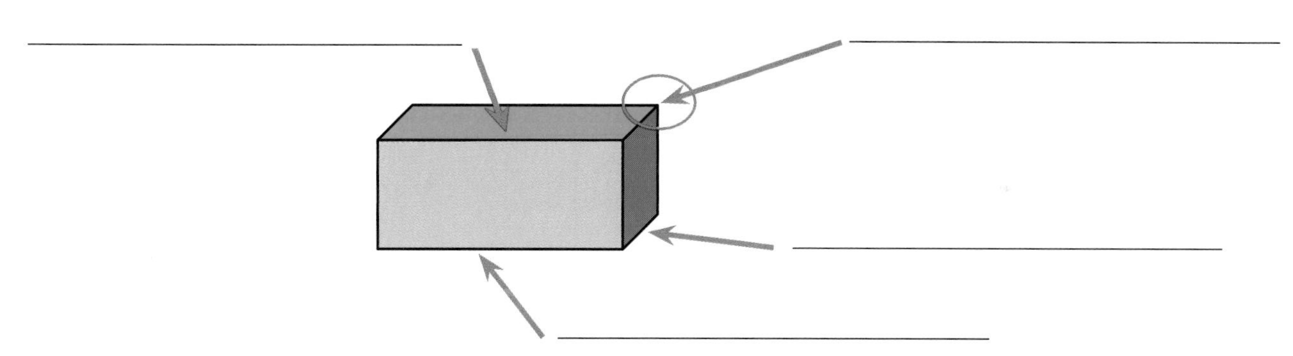

Use each model to fill in the blanks.

8		_____ faces _____ edges _____ vertices
9		_____ faces _____ edges _____ vertices
10		_____ faces _____ edges _____ vertices

Checkup

Write the letter of the example that matches each name.

1 ______ line

2 ______ line segment

3 ______ parallel lines

4 ______ perpendicular lines

A. 

B. 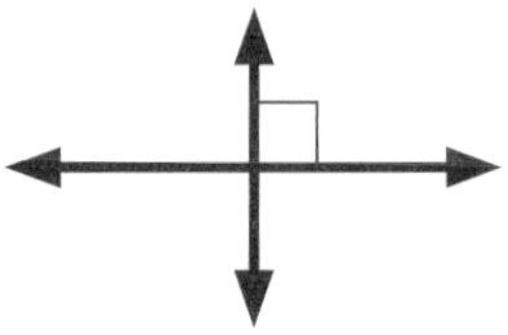

C. 

D. 

Fill in the chart.

	Polygon	Name	Number of Sides	Number of Angles
5	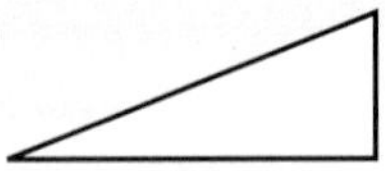		3	
6	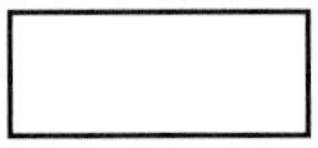	quadrilateral		
7	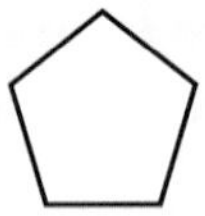	pentagon		

Decide if each sentence is TRUE or FALSE. Circle your answer.

8 A rhombus is a quadrilateral. TRUE FALSE

9 A trapezoid has 1 pair of parallel sides. TRUE FALSE

10 A rhombus has 4 right angles. TRUE FALSE

Write the letter of the example that best matches each name.

11 ______ cylinder

12 ______ pyramid

13 ______ sphere

14 ______ cone

15 ______ rectangular prism

16 ______ cube

A. 

B. 

C. 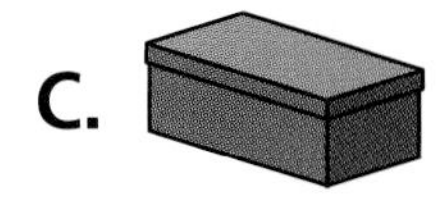

D. 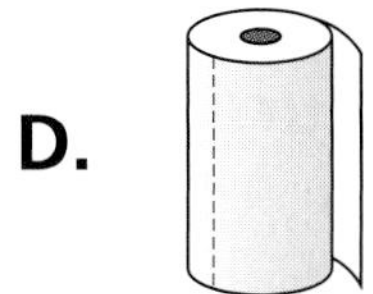

E. 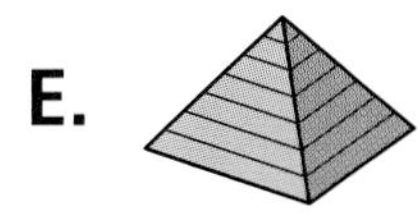

F.

Use each model to fill in the blanks.

17	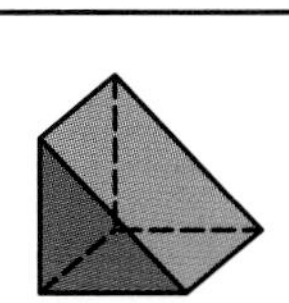	 ______ faces ______ edges ______ vertices
18	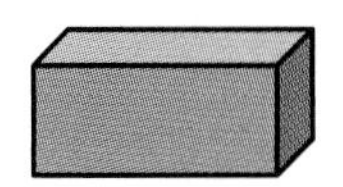	______ faces ______ edges ______ vertices
19	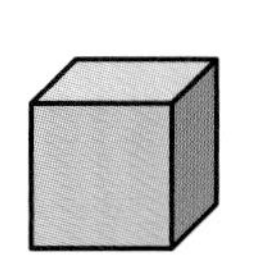	______ faces ______ edges ______ vertices

Lesson 32

Congruent and Similar

Congruent figures have the **same** shape and the **same** size.

▶ **Color the figure that is congruent to the first square.**

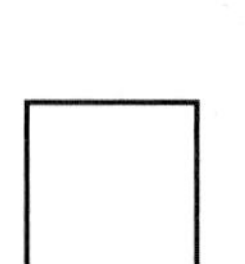
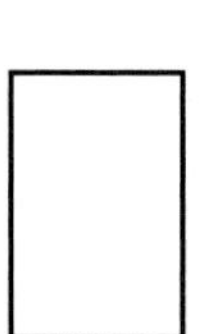

Similar figures have the **same** shape but are **different** sizes.

▶ **Circle the figure that is similar to the first triangle.**

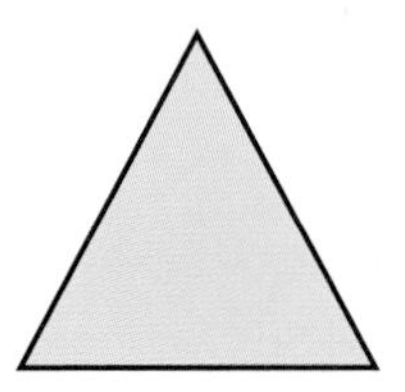
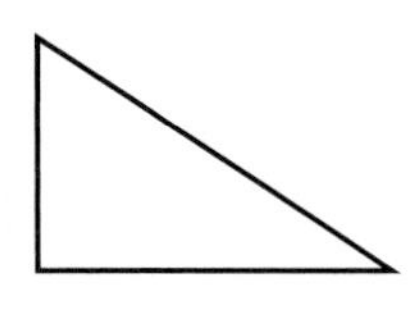
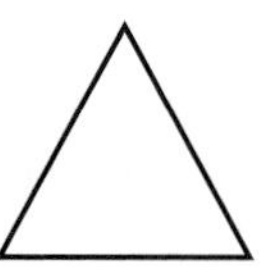

▶ **Look at the figures in the box.**

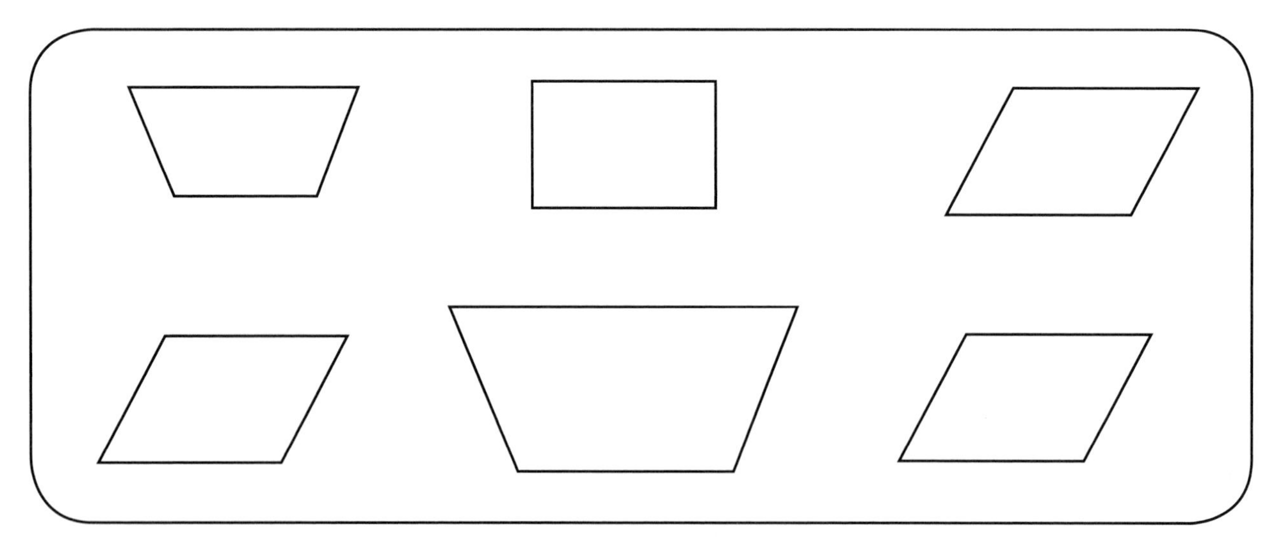

1 **Color the congruent figures.**

2 **Circle the similar figures.**

▶ Look at the figures in the box.

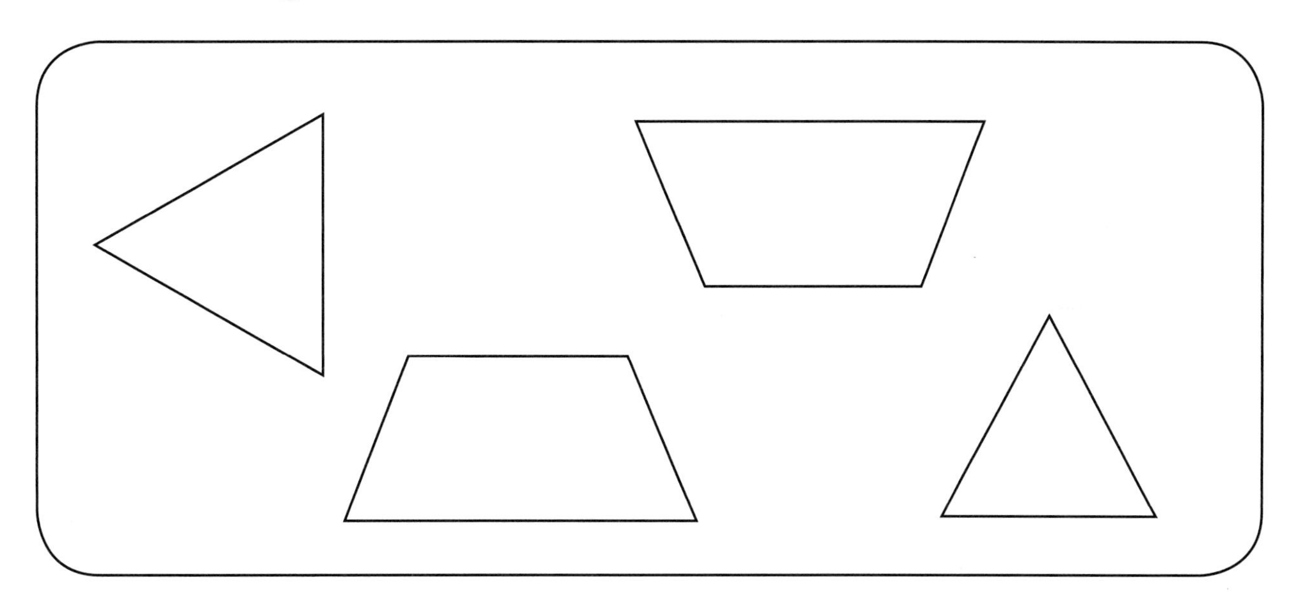

3 Color the congruent figures.

4 Circle the similar figures.

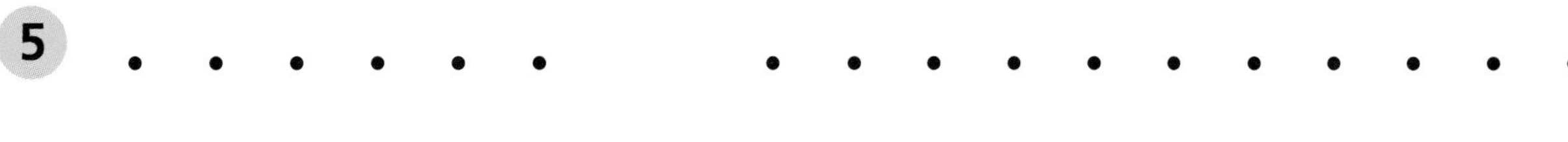

▶ Draw a congruent and a similar figure for each item.

5

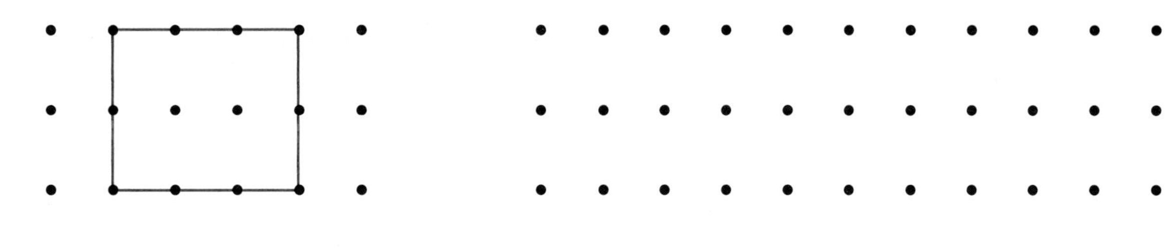

6

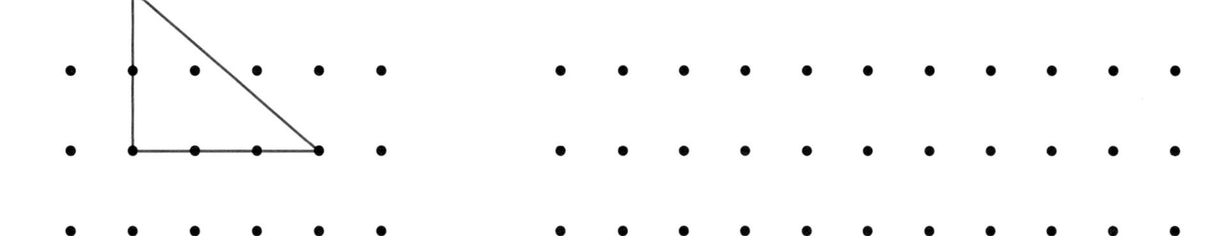

Lesson 33

Slides, Flips, and Turns

A figure can **slide**.

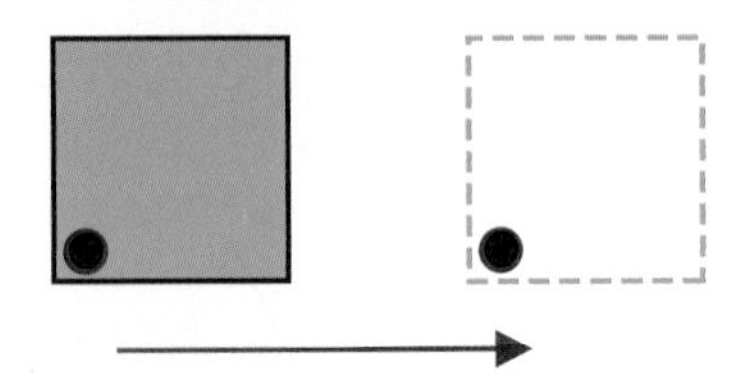

A figure can **turn**.

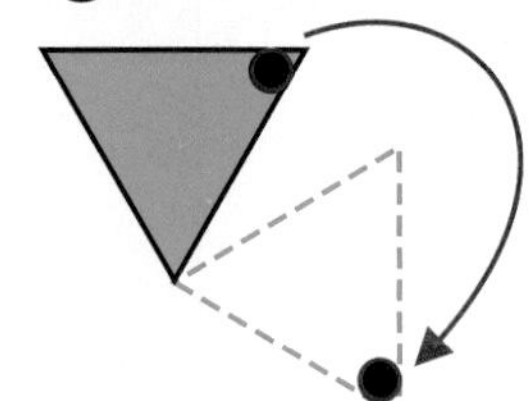

A figure can **flip**.

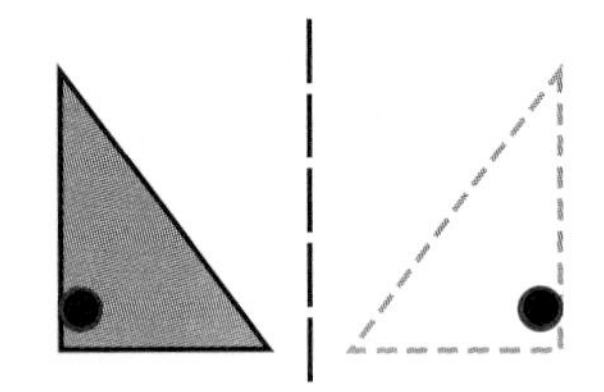

Draw lines to match each word to its example.

A. flip

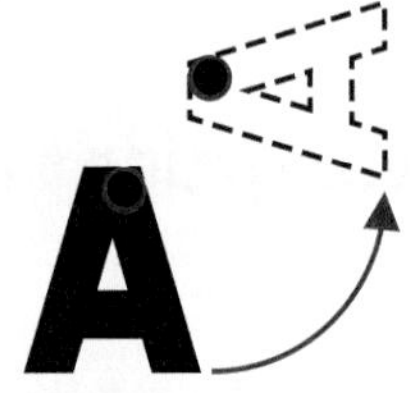

B. turn

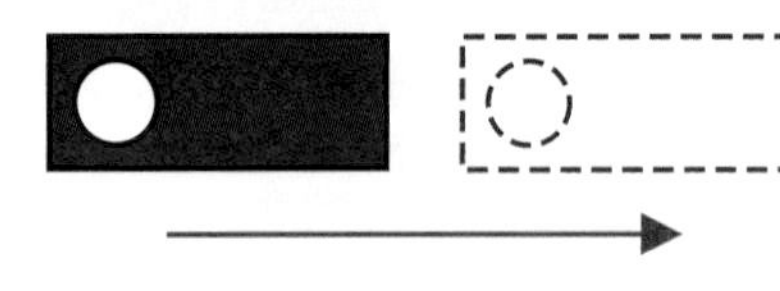

C. slide

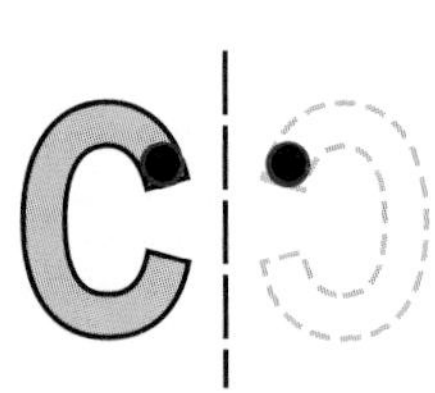

Draw the figure in its new place.

slide the rhombus

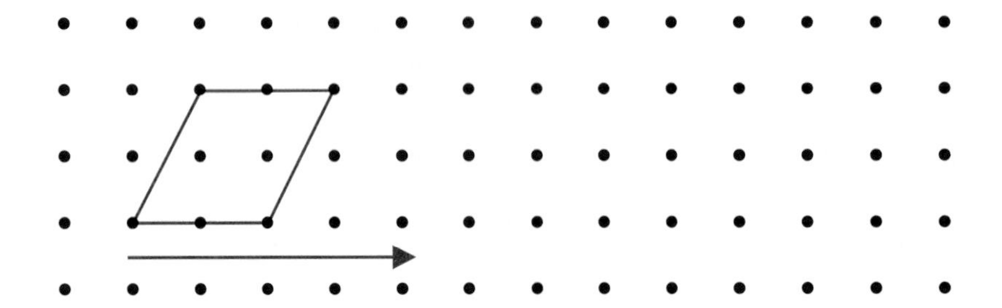

Use these words to tell how each figure moved.

slide flip turn

1 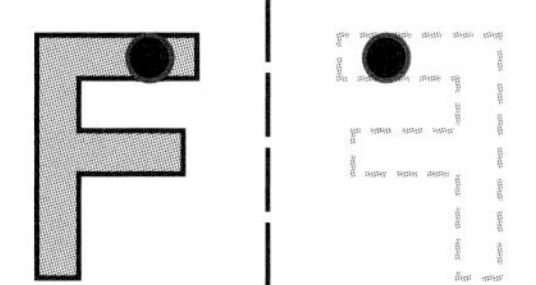

____

2 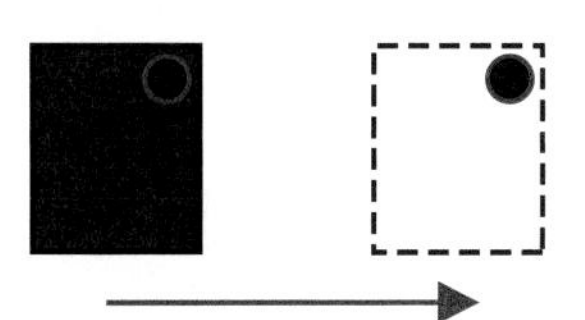

____

3 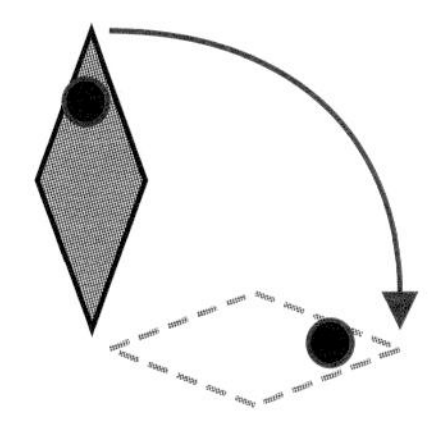

____

Draw the triangle in its new place.

4 Turn the triangle.

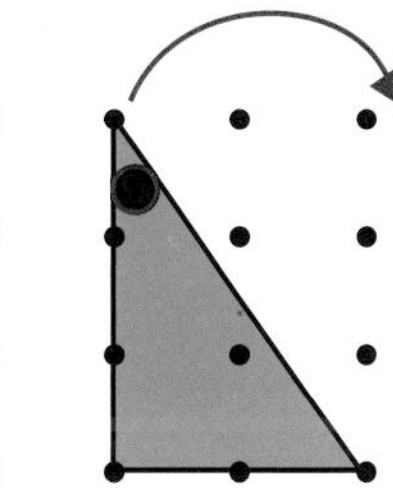
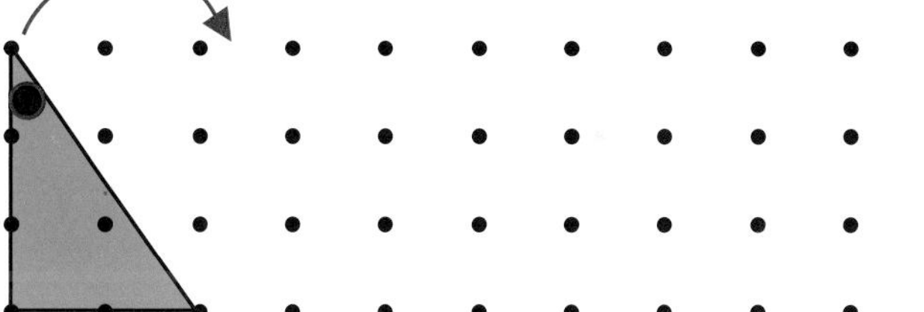

5 Flip the triangle.

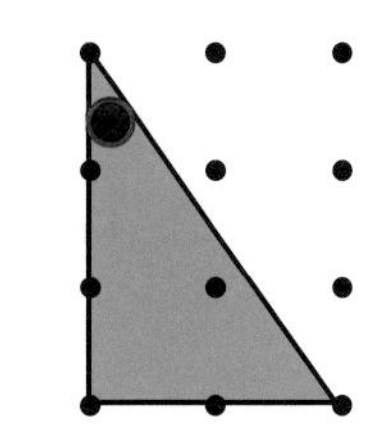
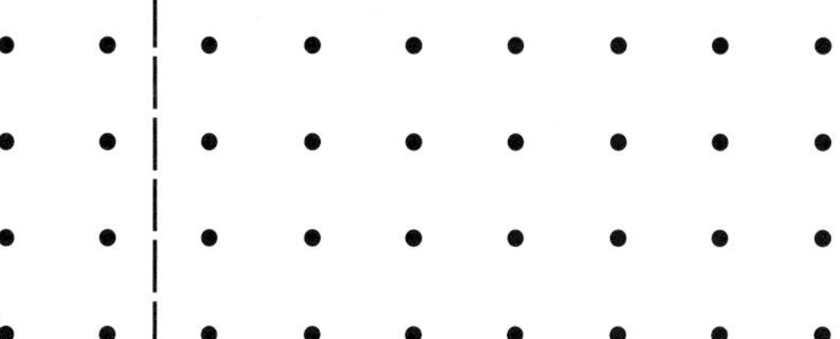

6 Slide the triangle.

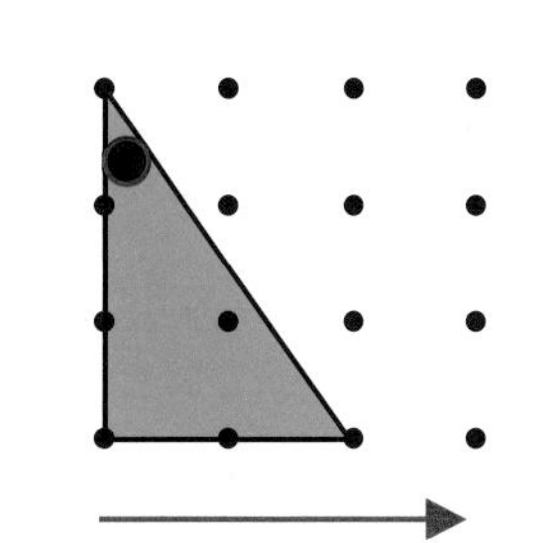

Lesson 34

Symmetry

A line of symmetry divides a figure into 2 matching halves.

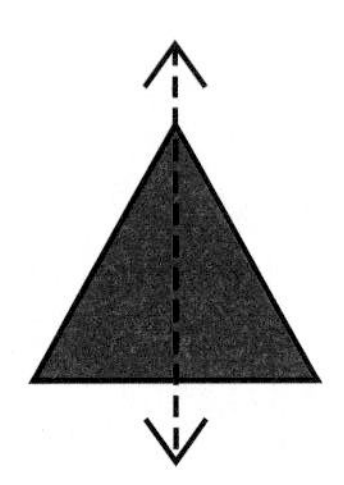

Some figures do not have a line of symmetry.

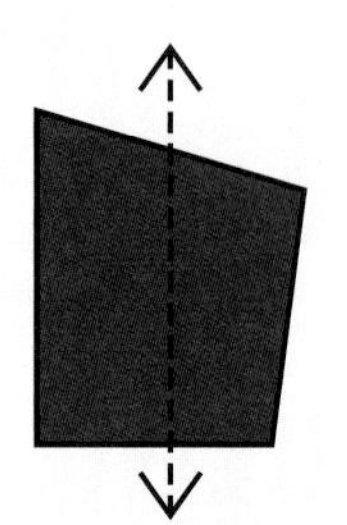

Circle the figures that show a line of symmetry.

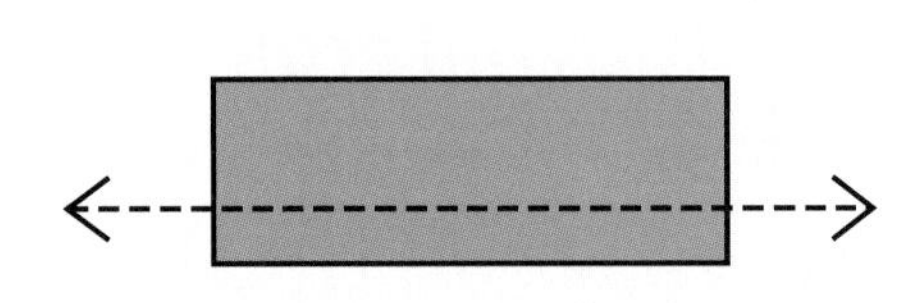

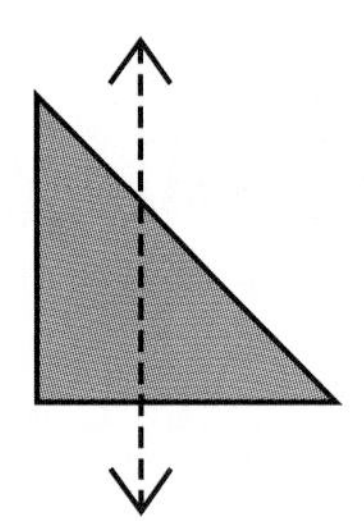

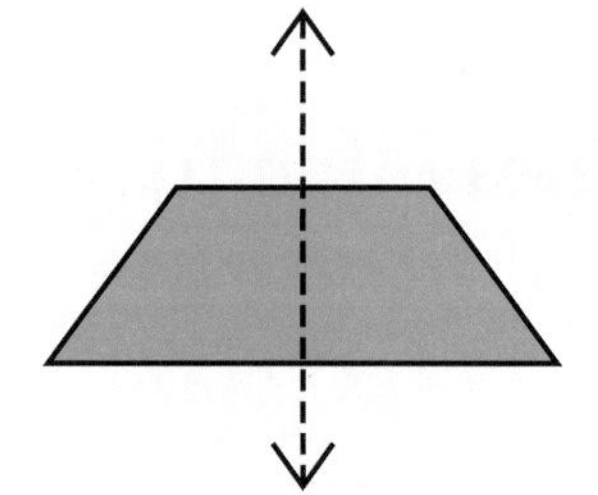

Draw all the lines of symmetry for this figure.

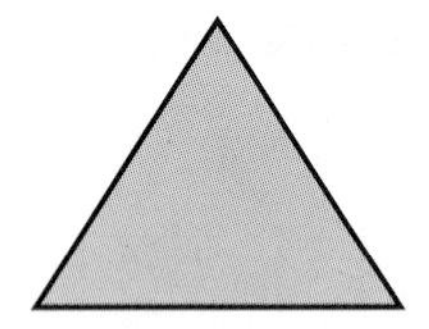

Draw the matching half across the line of symmetry.

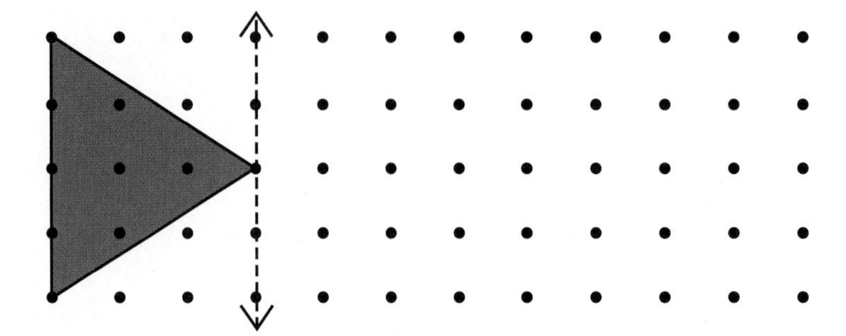

Circle the figures that show a line of symmetry.

1

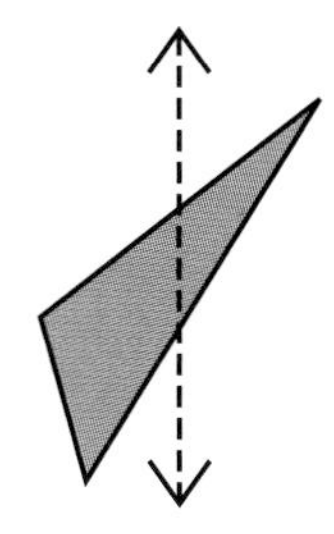

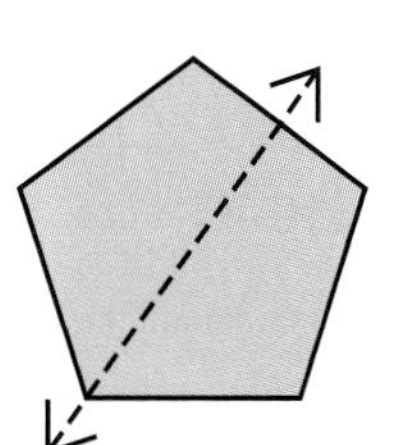

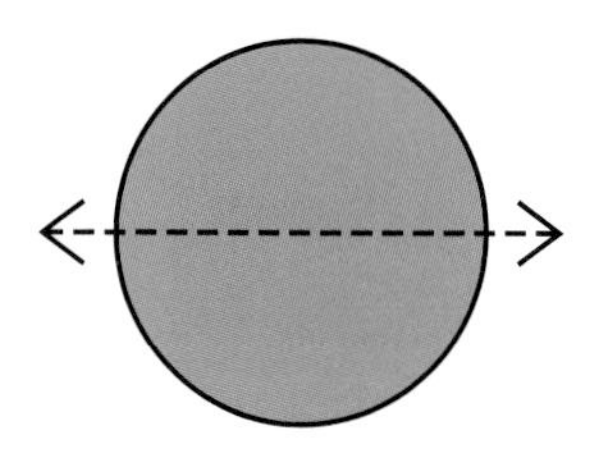

Draw all the lines of symmetry for each figure.

2

3

4

Draw the matching half across the line of symmetry.

5

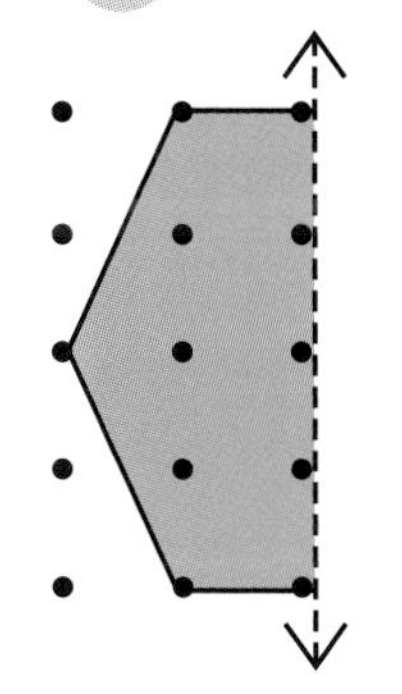

6

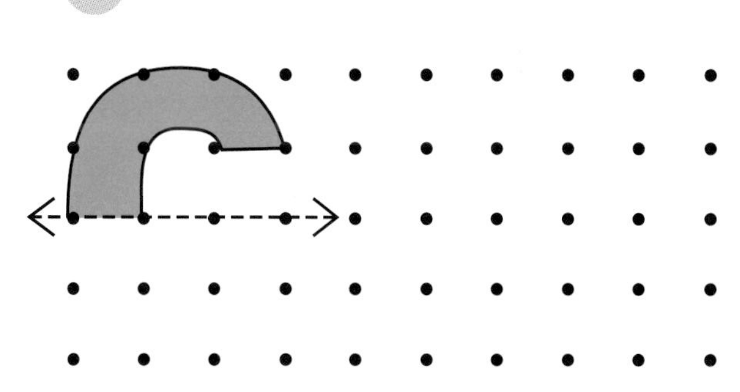

7

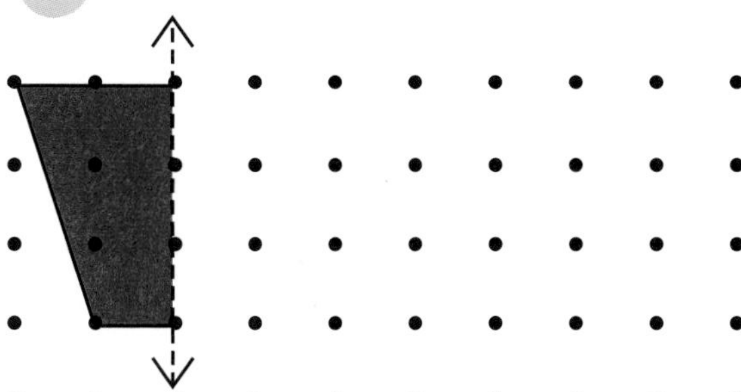

8

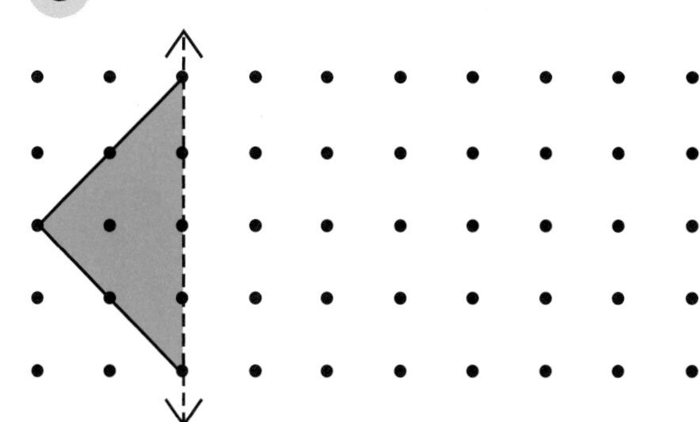

Checkup

▶ **Look at the figures in the box.**

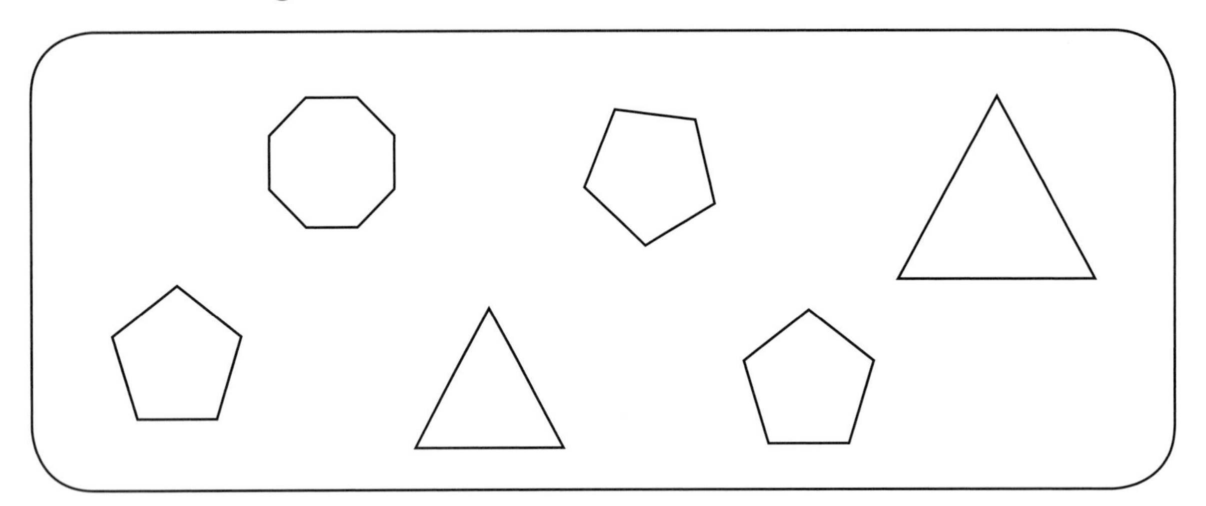

1 **Color the congruent figures.**

2 **Circle the similar figures.**

▶ **Draw a congruent and a similar figure for the rectangle.**

3

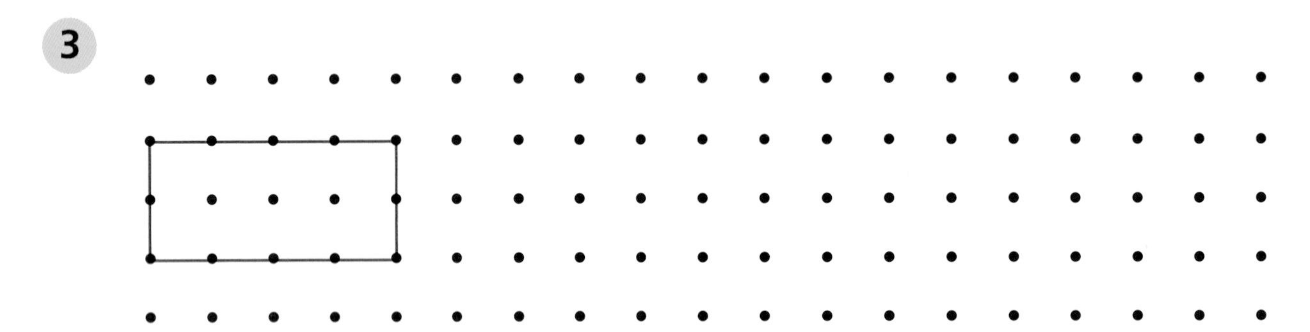

▶ **Use these words to tell how each figure moved.**

slide flip turn

4

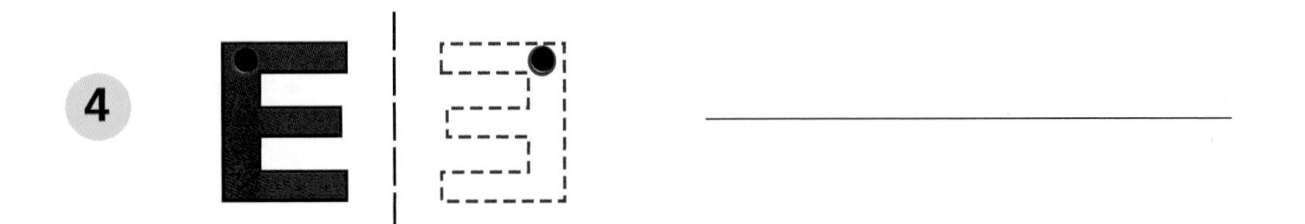

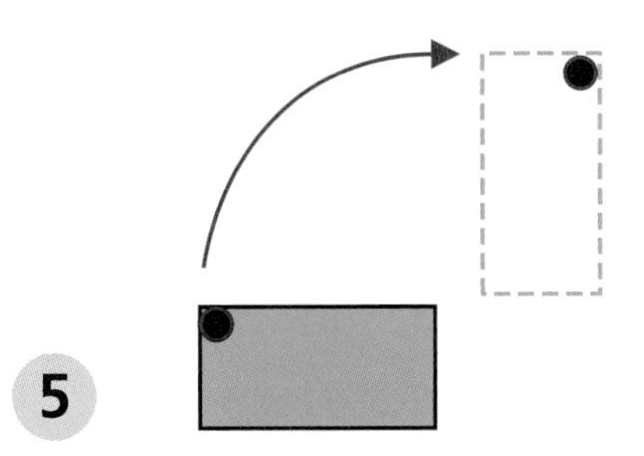

5 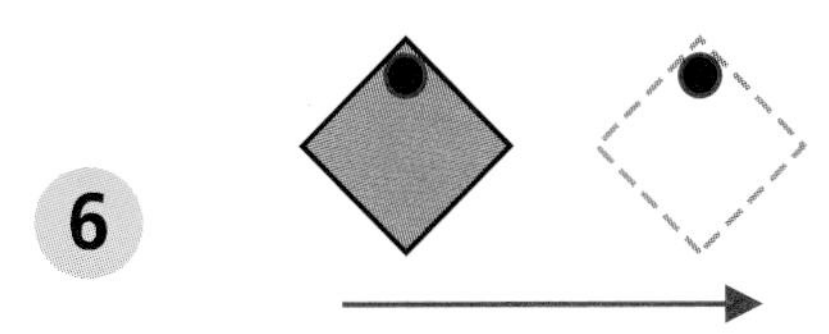 

6 

▶ **Draw the triangle in its new place.**

7 Turn the triangle.

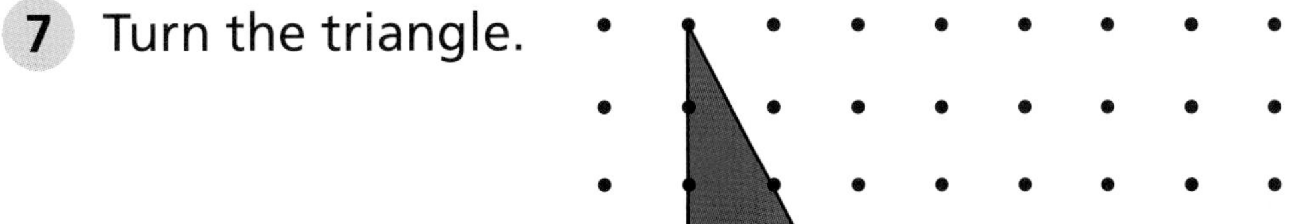

▶ **Circle the figures that show a line of symmetry.**

8

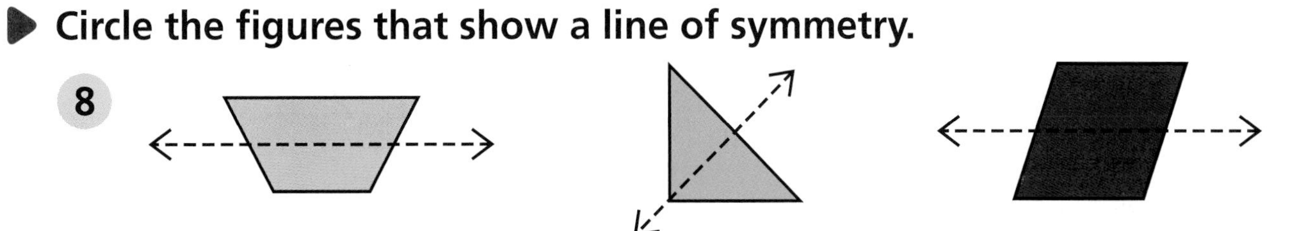

▶ **Draw all the lines of symmetry for each figure.**

9 10

▶ **Draw the matching half across the line of symmetry.**

11 12

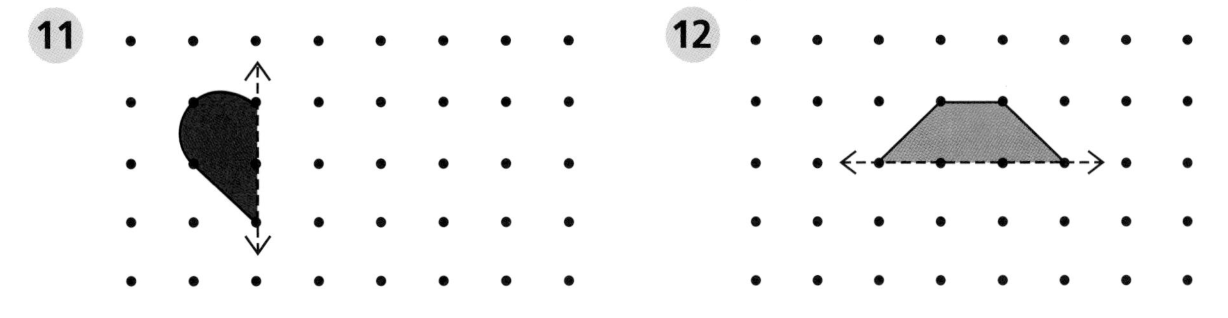

Lesson 35

Perimeter in Customary Units

Perimeter (P) is the distance around a figure. Perimeter can be measured in customary units, such as inches and feet. You can find the perimeter of a figure by adding the lengths of its sides.

▶ **Use a ruler to measure the perimeter of this figure.**

P = 3 in. + 2 in. + 3 in. + 2 in. =

_____ in.

▶ **Find the perimeter of this figure.**

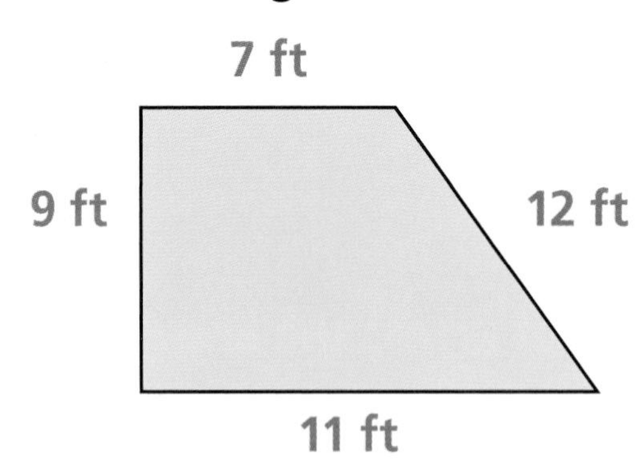

P = _____ ft + _____ ft + _____ ft + _____ ft = _____ ft

Use a ruler to measure the perimeter of this figure.

1

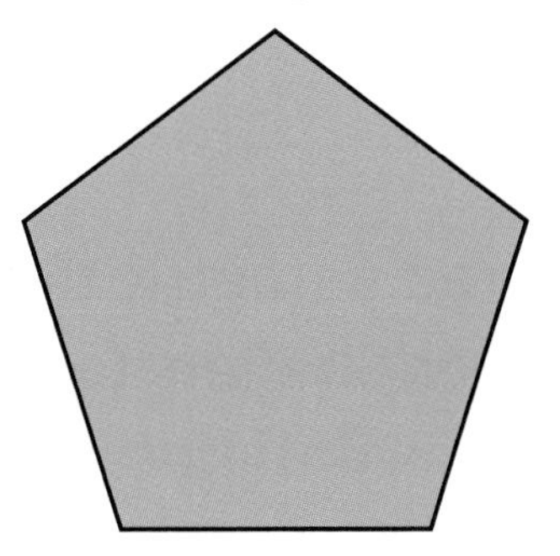

P = ______ in. + ______ in. + ______ in. + ______ in. + ______ in. = ______ in.

Find the perimeter of each figure.

2

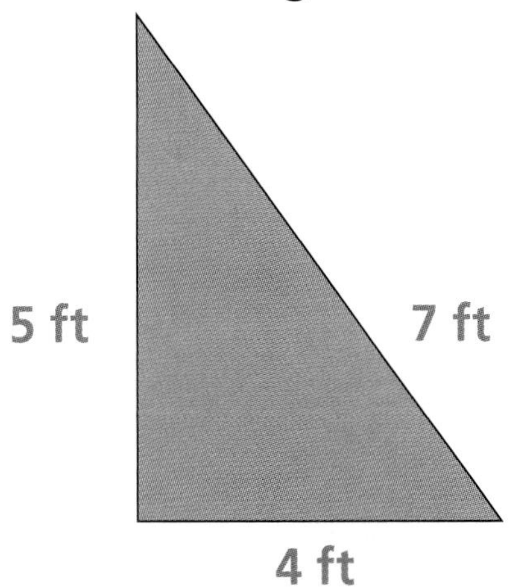

P = ______ ft + ______ ft + ______ ft = ______ ft

3

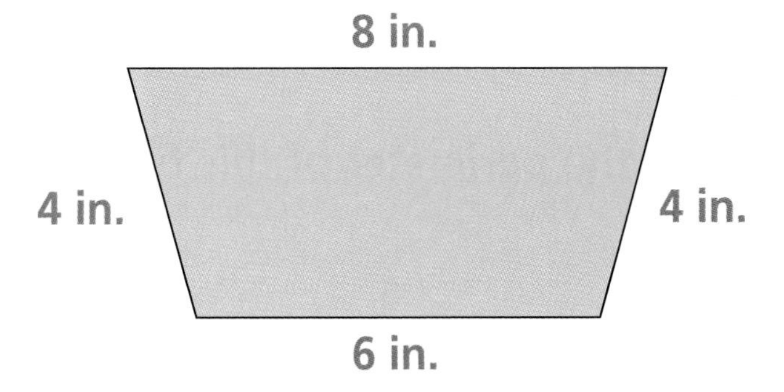

P = ______ in. + ______ in. + ______ in. + ______ in. = ______ in.

Lesson 36

Perimeter in Metric Units

Perimeter (P) can also be measured in metric units, such as **meters** and **centimeters**. There are 100 centimeters (cm) in a meter (m). There are 10 **millimeters** (mm) in a centimeter.

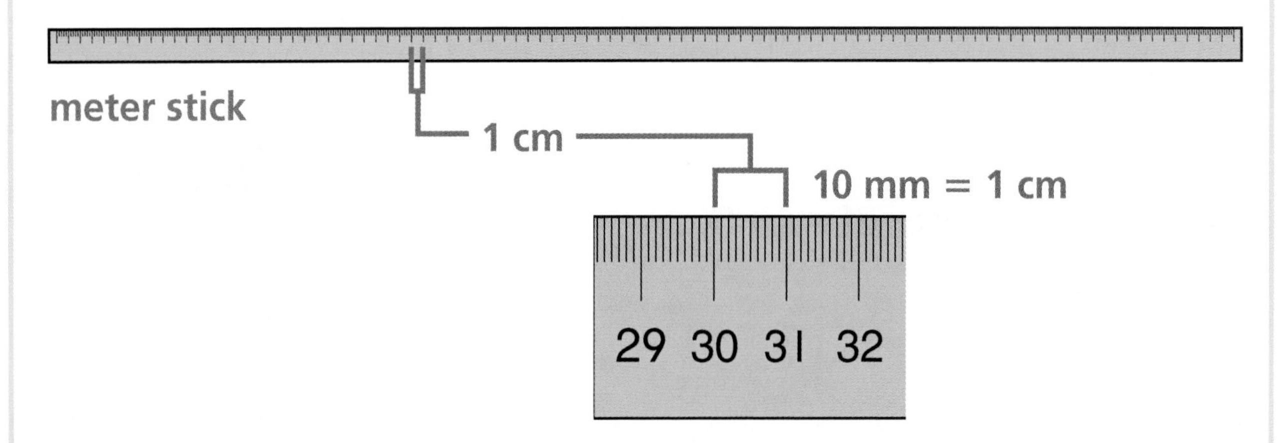

▶ **Use a ruler to measure the perimeter of this figure to the nearest millimeter.**

P = 15 mm + 15 mm + 15 mm + 15 mm

= ______ mm

▶ **Use a ruler to measure the perimeter of this figure to the nearest centimeter.**

1

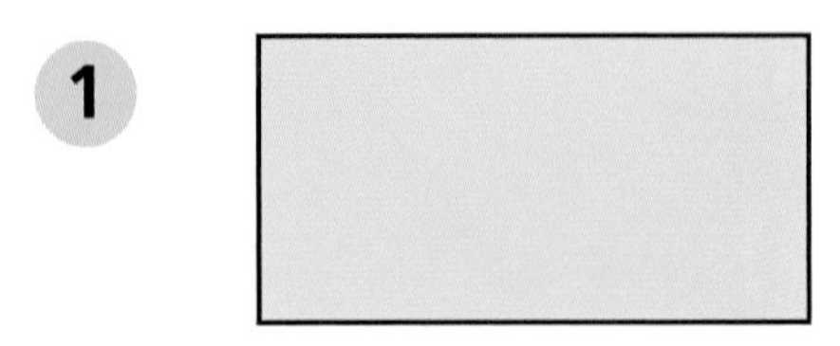

P = ______ cm + ______ cm + ______ cm + ______ cm = ______ cm

 **Find the perimeter of each figure.**

2

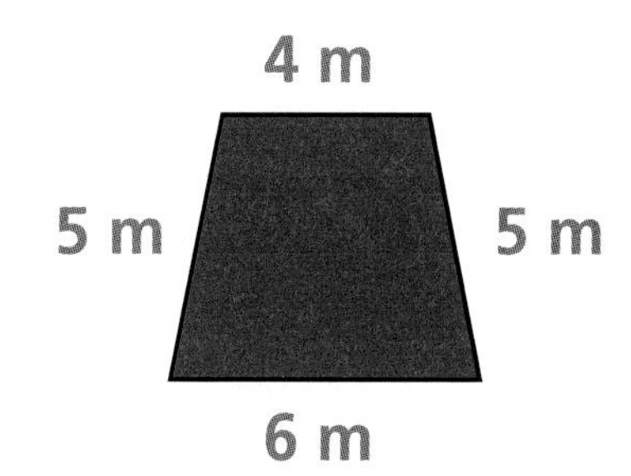

P = ______ m + ______ m + ______ m + ______ m = ______ m

3

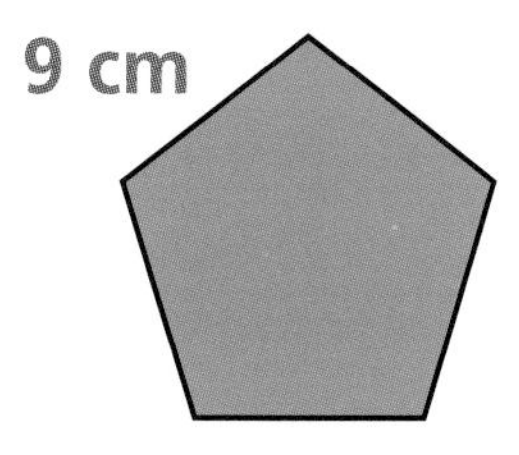

Each side is 9 cm long.

P = ______ cm + ______ cm + ______ cm + ______ cm +

______ cm = ______ cm

4

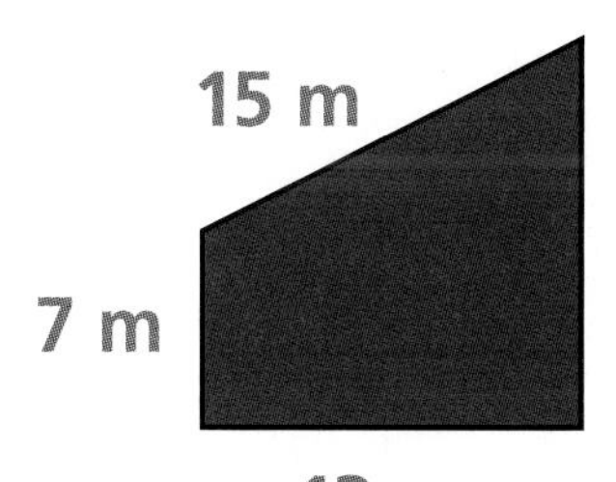

P = ______ m + ______ m + ______ m + ______ m = ______ m

5

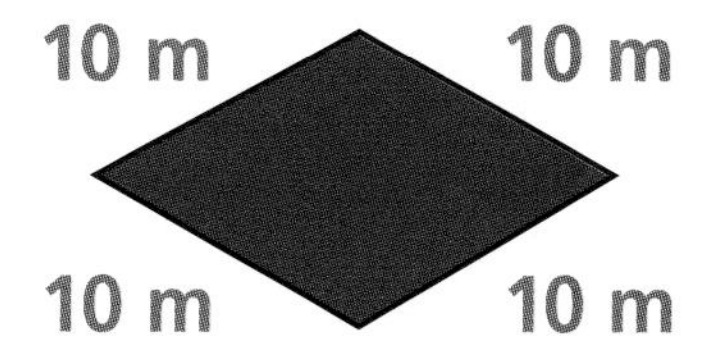

P = ______ m + ______ m + ______ m + ______ m = ______ m

Lesson 37

Explore Area

Area (A) is the number of square units that cover a figure.

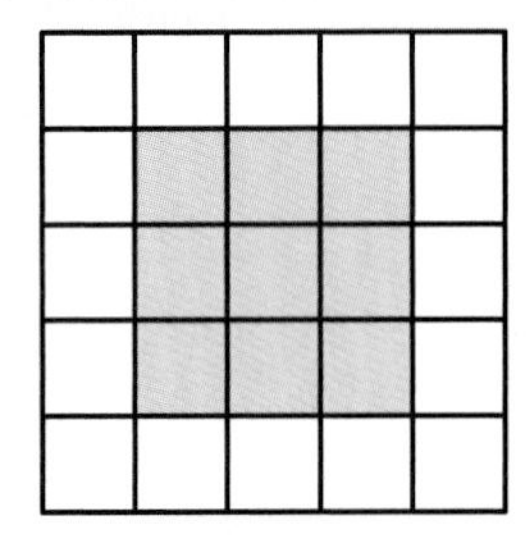

A = 9 square units

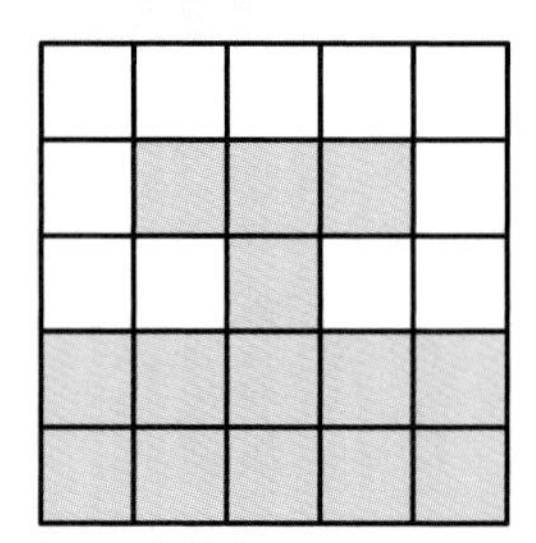

A = 14 square units

▶ **Find the area of each figure.**

A.

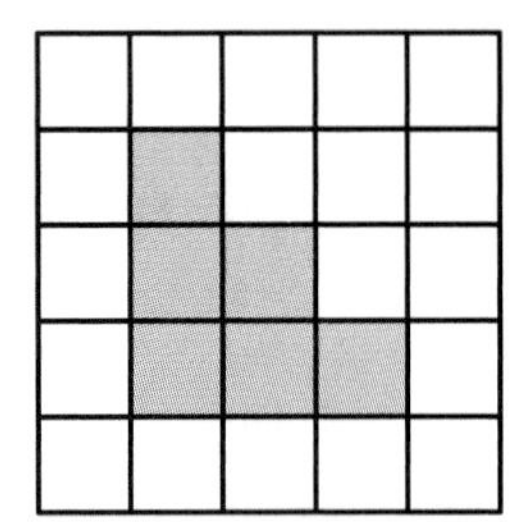

A = 6 square units

B.

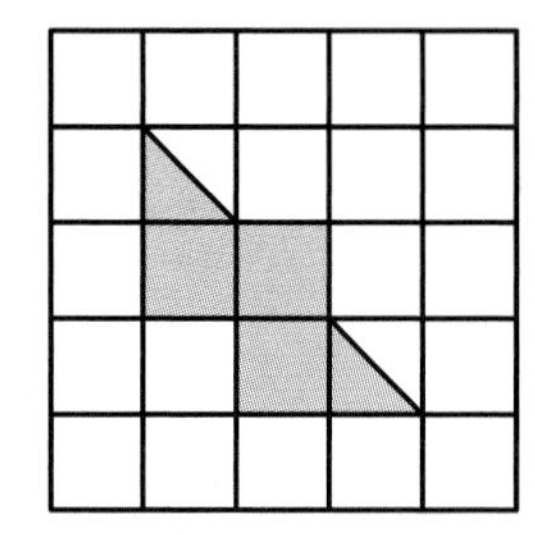

A = ______ square units

▶ **Find the area of each figure.**

1

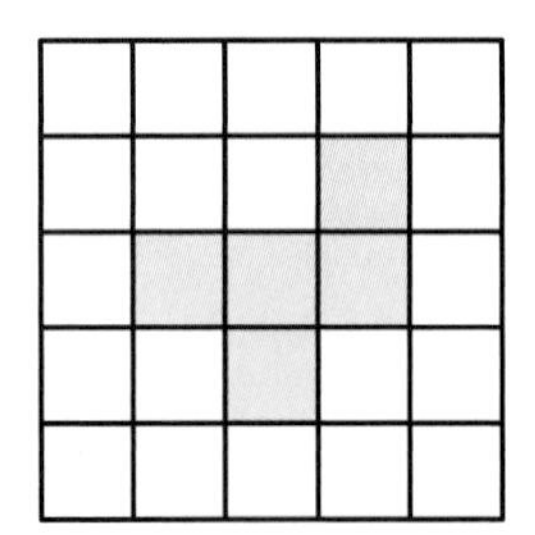

A = ______ square units

2

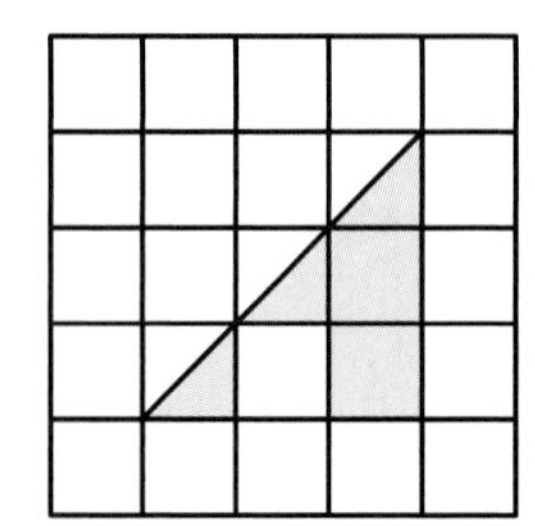

A = ______ square units

3

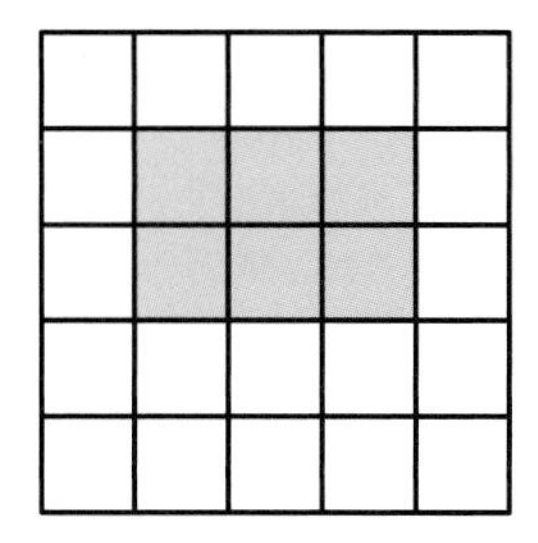

A = ______ square units

4

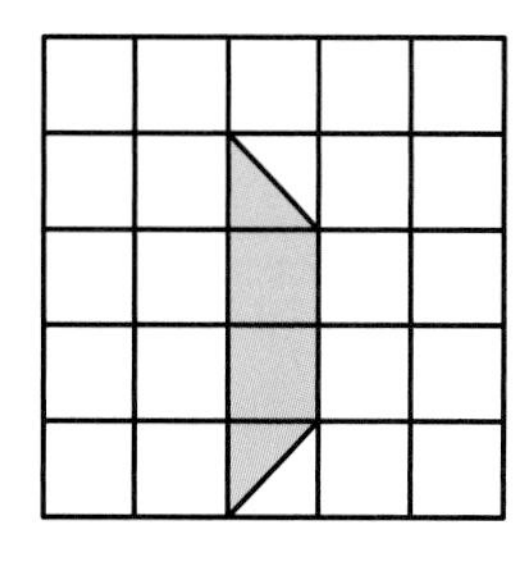

A = ______ square units

5

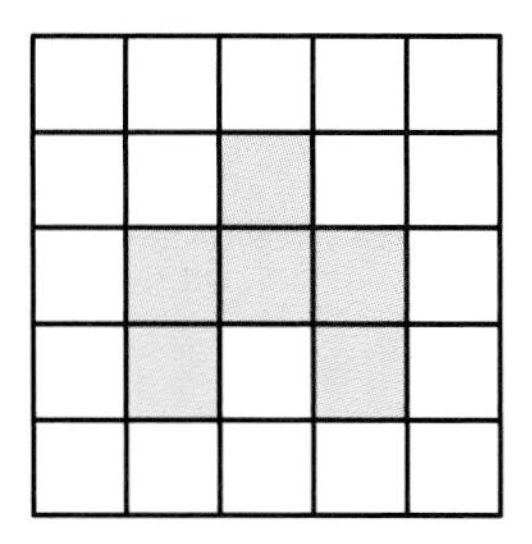

A = ______ square units

6

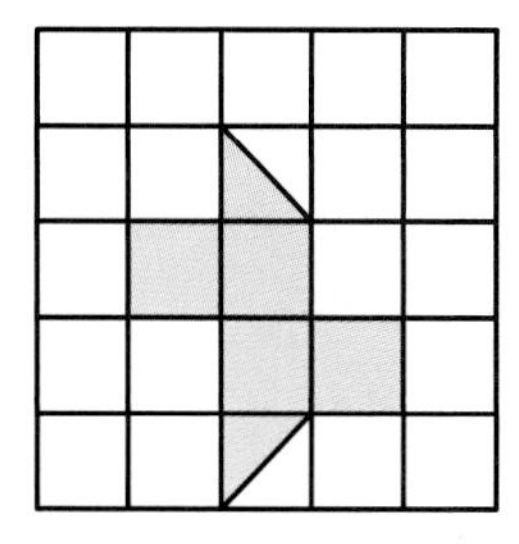

A = ______ square units

7

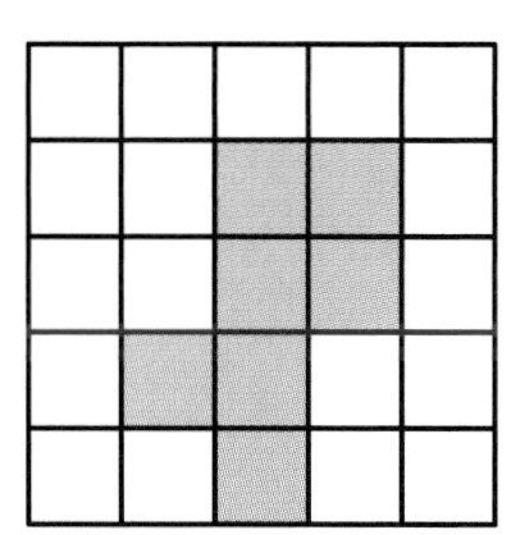

A = ______ square units

8

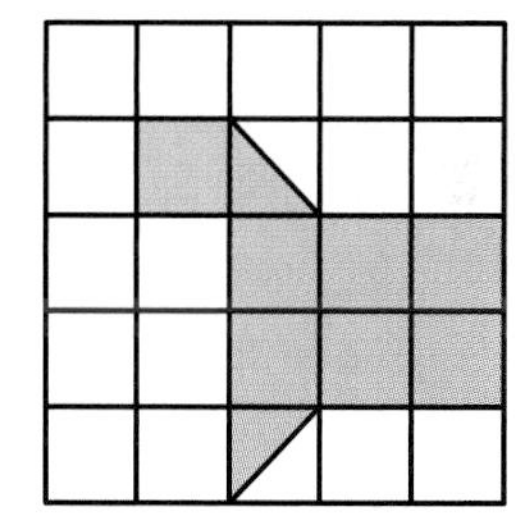

A = ______ square units

Lesson 38

Explore Volume

Volume (V) is the number of cubes that it takes to fill a solid figure.

Find the volume of each solid figure.

A.

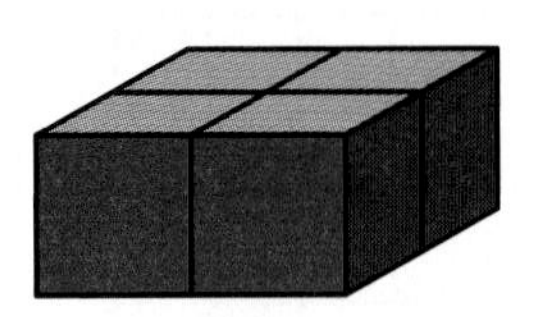

V = 4 cubic units

B.

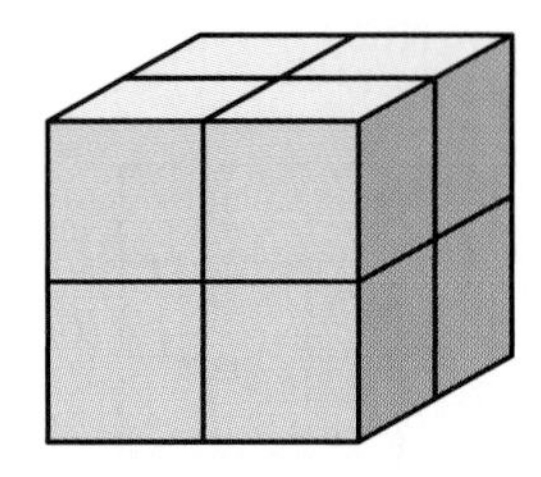

V = ______ cubic units

C.

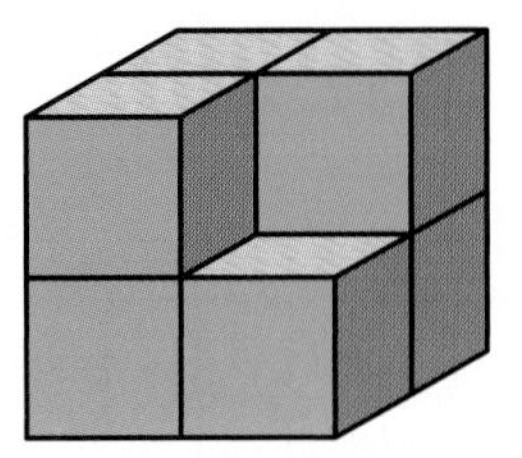

V = ______ cubic units

D.

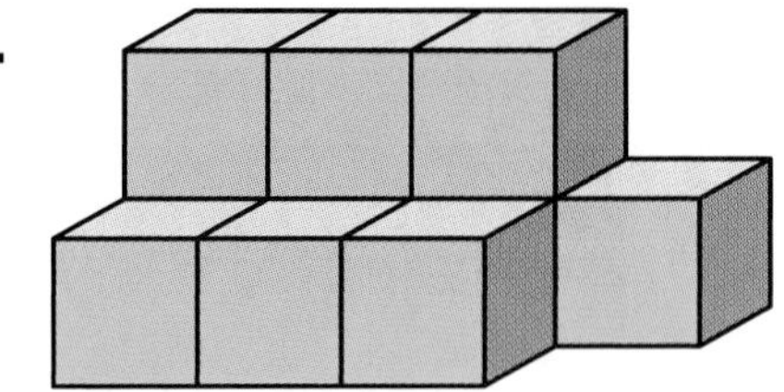

V = ______ cubic units

Find the volume of each solid figure.

1

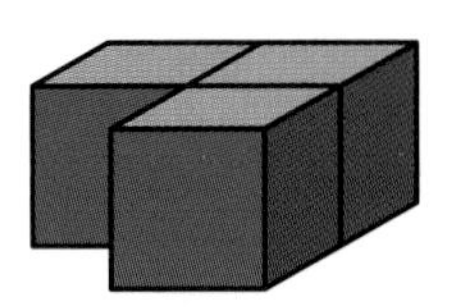

V = ______ cubic units

2

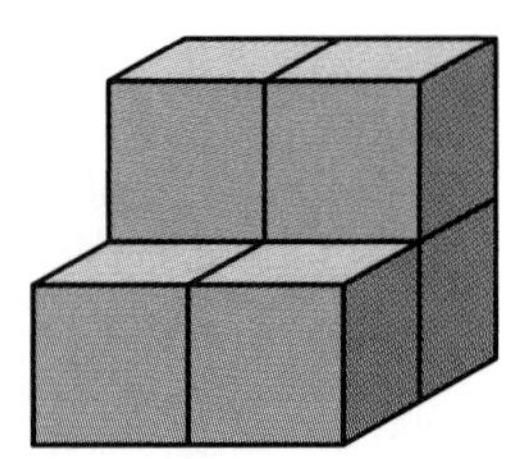

V = ______ cubic units

3

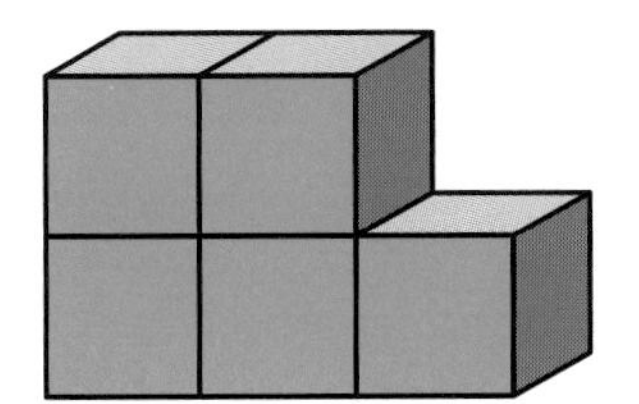

V = ______ cubic units

4

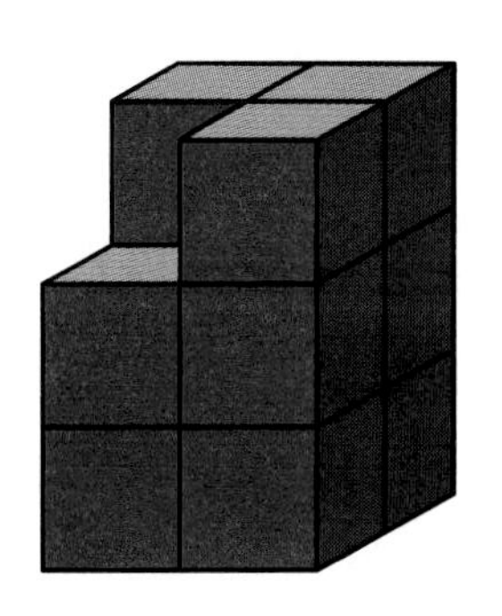

V = ______ cubic units

5

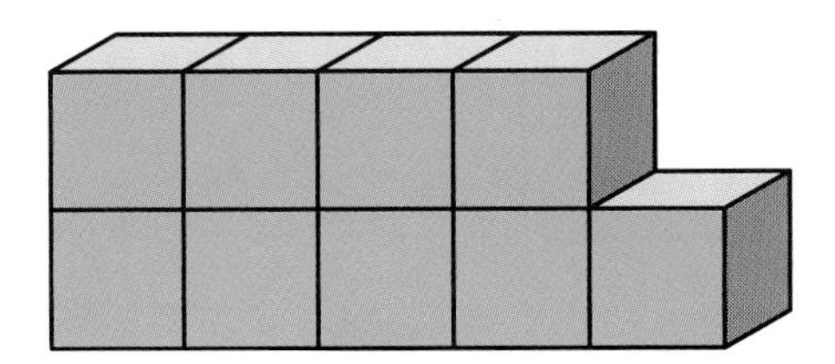

V = ______ cubic units

6

V = ______ cubic units

7

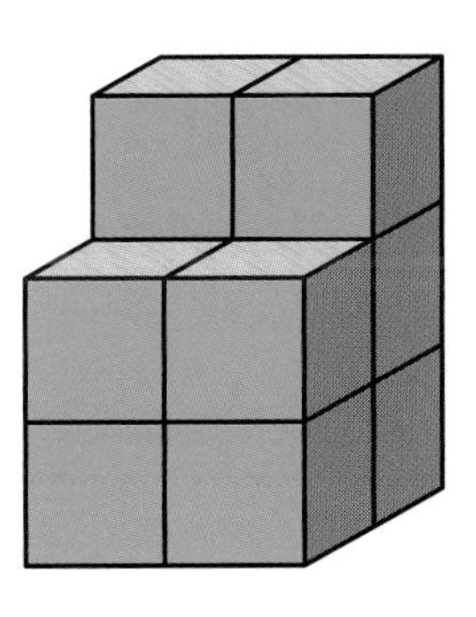

V = ______ cubic units

8

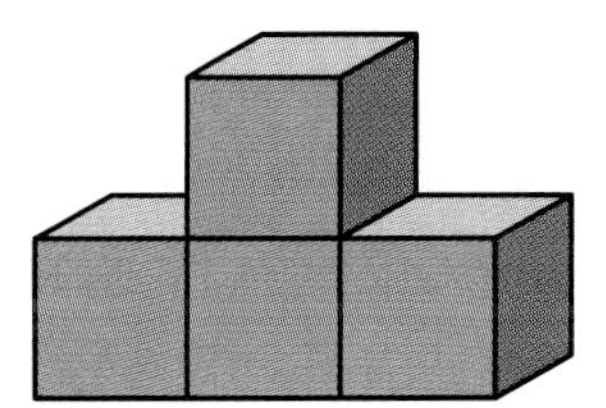

V = ______ cubic units

Lesson 39

Measure Capacity in Customary Units

Capacity is the amount of liquid a container can hold.

Customary Units of Measure
2 cups (c) = 1 pint (pt)
4 cups (c) = 2 pints (pt) = 1 quart (qt)
4 quarts (qt) = 1 gallon (gal)

1 cup (c) 1 pint (pt) 1 quart (qt) 1 gallon (gal)

Circle the capacity of each container.

A.

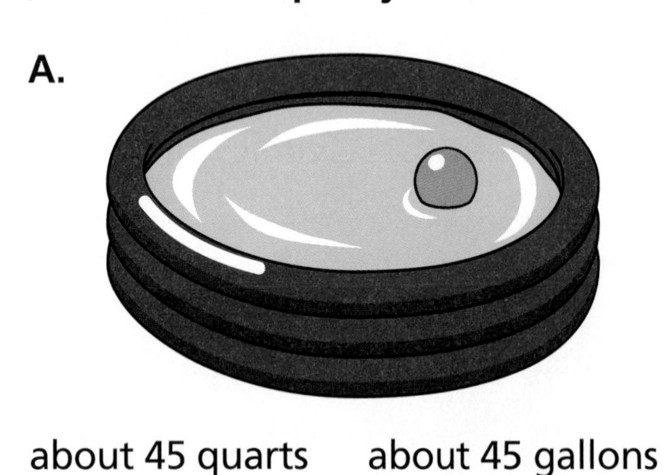

about 45 quarts about 45 gallons

B.

about 1 cup about 1 pint

Draw lines to match each word to its best example.

1 pint

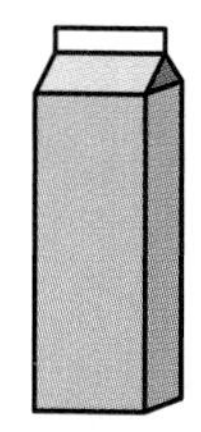

2 gallon

3 cup

4 quart

Circle the capacity of each container.

5

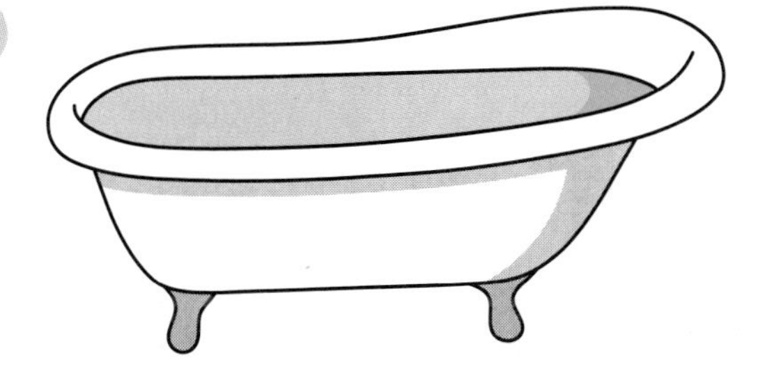

about 40 quarts | about 40 gallons

6

about $\frac{1}{2}$ gallon | about $\frac{1}{2}$ pint

7

about 12 pints | about 12 gallons

8

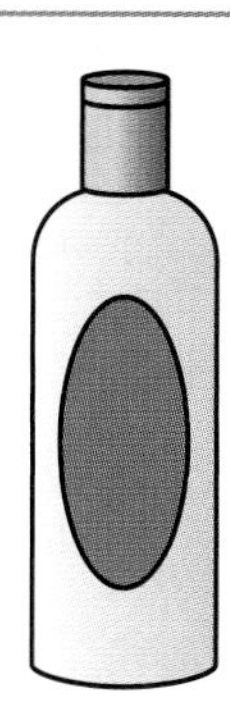

about 3 cups | about 3 gallons

Lesson 40

Measure Capacity in Metric Units

You can also use metric units to measure capacity. Metric units for capacity are **milliliters** (mL) and **liters** (L).

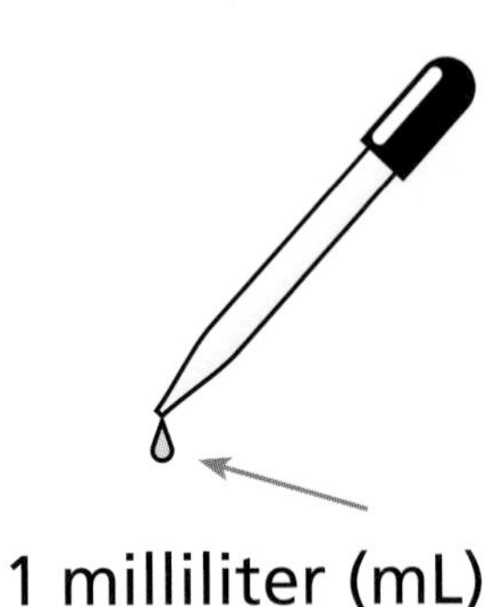

1 milliliter (mL)

1 liter (L)

Circle the capacity of each container.

1

about 750 milliliters | about 750 liters

2

about 6 milliliters | about 6 liters

3

about 240 milliliters | about 240 liters

4

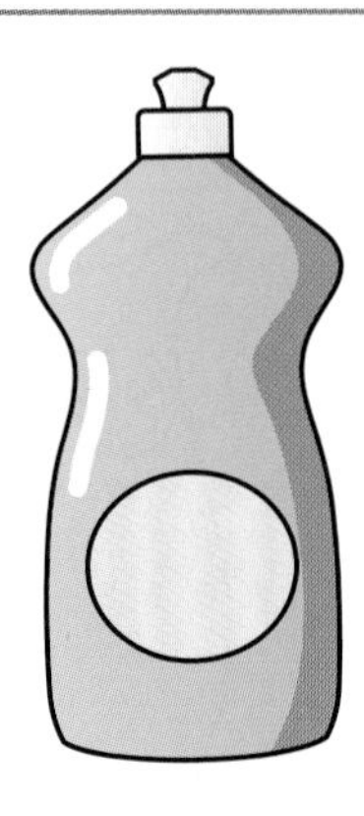

about 1 milliliter | about 1 liter

Circle the unit of measure you would use to measure the capacity of each container.

5 kitchen sink — liter or milliliter

6 juice box — liter or milliliter

7 pitcher of water — liter or milliliter

8 medicine bottle — liter or milliliter

Draw lines to match each measurement to its best example.

9 200 L

10 300 L

11 5 mL

12 4 L

Checkup

Use a ruler to measure the perimeter of each figure.

1

P = ______ in. + ______ in. + ______ in. + ______ in. = ______ in.

2

P = ______ cm + ______ cm + ______ cm = ______ cm

Find the perimeter of the figure.

3

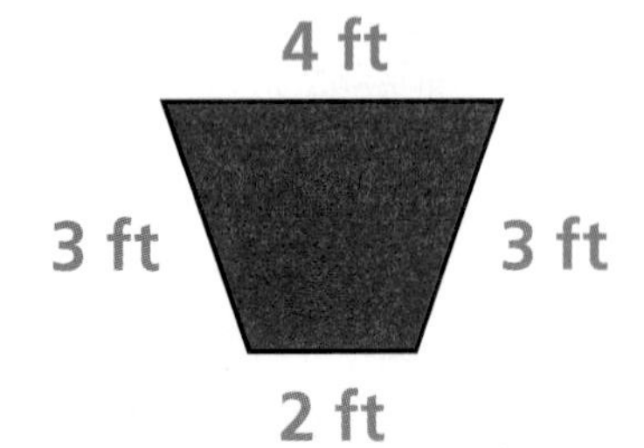

P = ______ ft + ______ ft + ______ ft + ______ ft = ______ ft

Find the area of each figure.

4

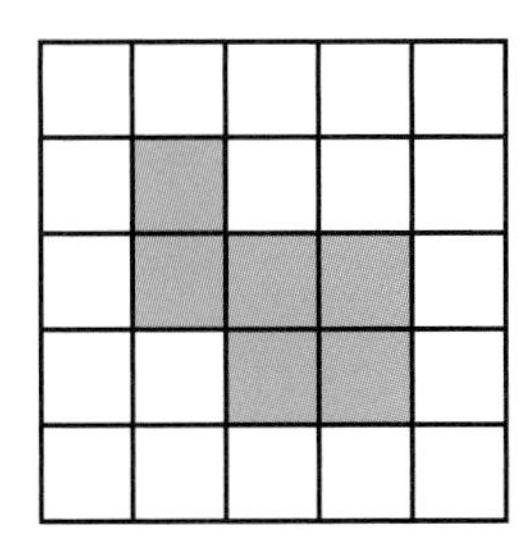

A = ______ square units

5

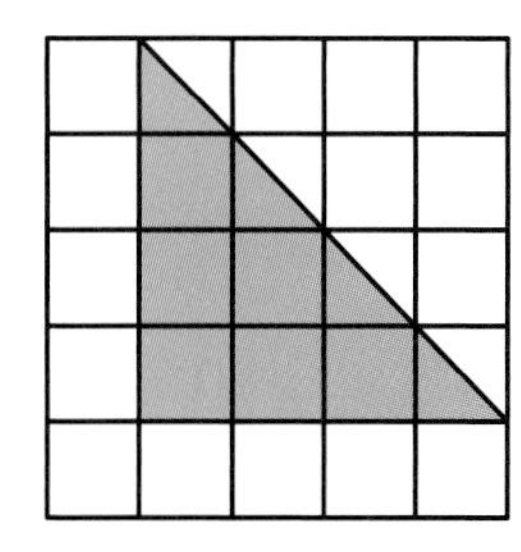

A = ______ square units

Find the volume of each solid figure.

6

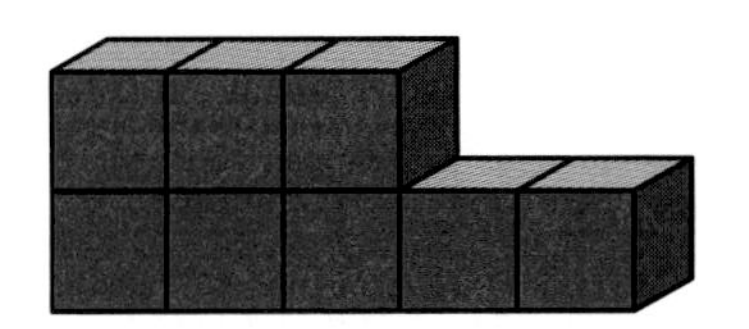

V = ______ cubic units

7

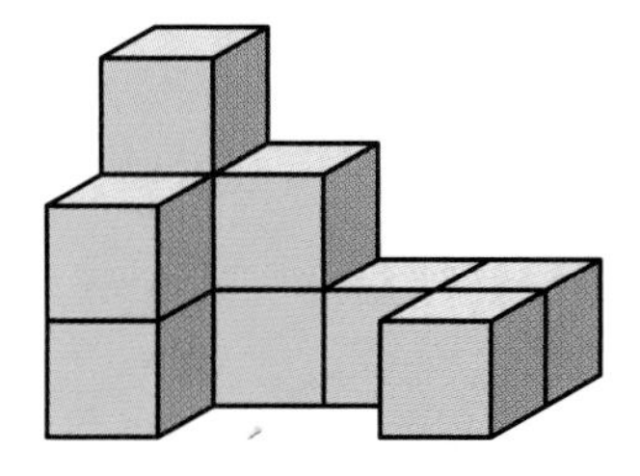

V = ______ cubic units

Circle the capacity of each container.

8

about 100 ounces | about 100 pints

9

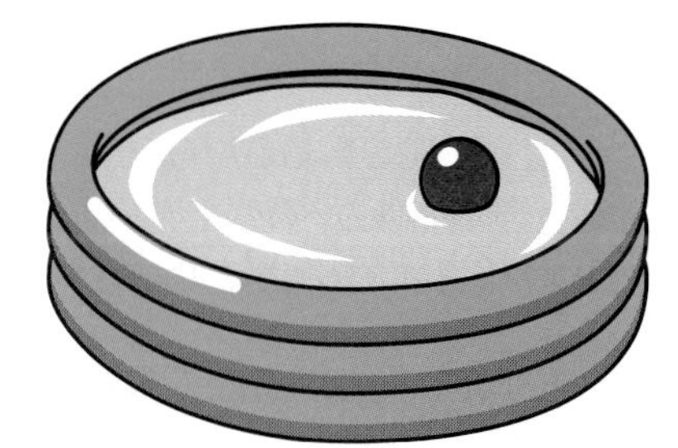

about 40 cups | about 40 gallons

10

about 4 milliliters | about 4 liters

11

about 25 milliliters | about 25 liters

Lesson 41

Weight in Customary Units

You can use customary units to measure an object's **weight**. The customary units for weight are **ounces** (oz) and **pounds** (lb). There are 16 ounces in one pound.

Circle the weight of each object.

about 1 ounce   about 1 pound

about 1 ounce   about 1 pound

Circle the unit you would use to measure the weight of each object.

dictionary

ounces or pounds

seashell

ounces or pounds

Fill in the blank with the missing measure.

2 lb = ______ oz

There are 16 ounces in 1 pound. $2 \times 16 =$ ______ oz

There are 32 ounces in 2 pounds.

Circle the weight of each object.

1

about 10 ounces   about 10 pounds

2

about 9 ounces   about 9 pounds

3

about 30 ounces | about 30 pounds

4

about 48 ounces | about 48 pounds

Draw lines to match each measurement to its best example.

5 50 ounces

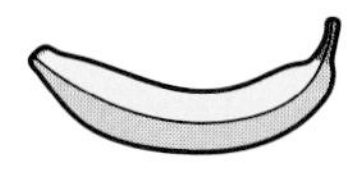

6 50 pounds

7 8 ounces

8 8 pounds

Circle the unit you would use to measure the weight of each object.

9 bike  ounces or pounds

10 carrot 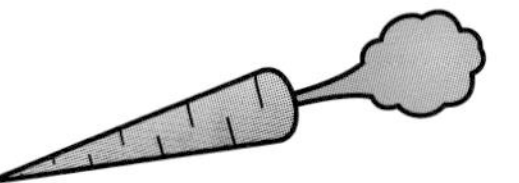 ounces or pounds

Fill in the blank of each missing measure.

11 3 lb = ______ oz

12 32 oz = ______ lb

Lesson 42

Mass in Metric Units

You can use metric units to measure an object's mass. The metric units for mass are **grams** (g) and **kilograms** (kg). There are 1,000 grams in 1 kilogram.

Circle the mass of each object.

about 1 gram   about 1 kilogram

about 1 gram   about 1 kilogram

Fill in the blank of the missing measure.

3 kg = ______ g

Circle the unit you would use to measure the object's mass.

bowl

grams or kilograms

Circle the mass of each object.

1

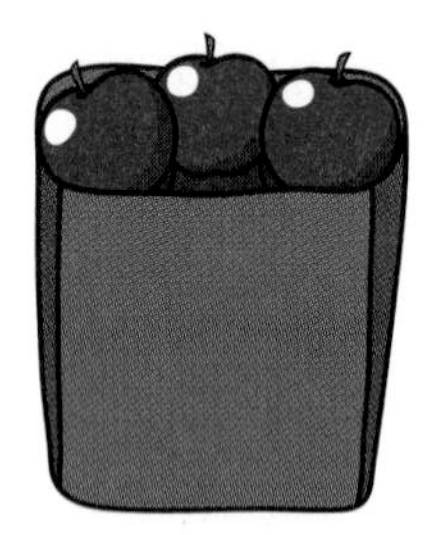

about 2 grams   about 2 kilograms

2

about 25 grams   about 25 kilograms

3

about 10 grams   about 10 kilograms

4

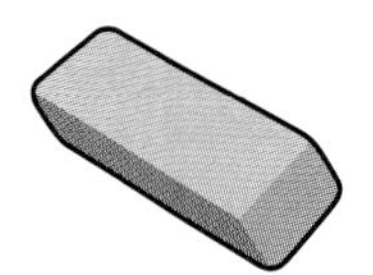

about 10 grams   about 10 kilograms

▶ **Draw lines to match each measurement of mass to its best example.**

5 100 grams

6 400 kilograms

7 5 grams

8 25 kilograms

▶ **Fill in the blank of the missing measure.**

9 $\frac{1}{2}$ kg = ______ g

10 2,000 g = ______ kg

▶ **Circle the unit you would use to measure each object's mass.**

11 brick 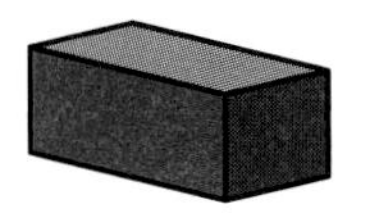 grams or kilograms

12 banana 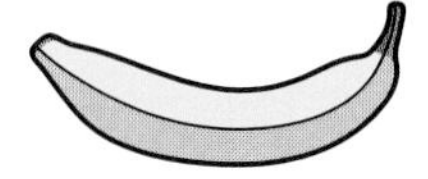 grams or kilograms

Explore Temperature

You can use a thermometer to measure something's **temperature**, or how hot or cold something is.

You can use customary or metric units to measure temperature. The customary unit is **degrees Fahrenheit** (°F). The metric unit is **degrees Celsius** (°C).

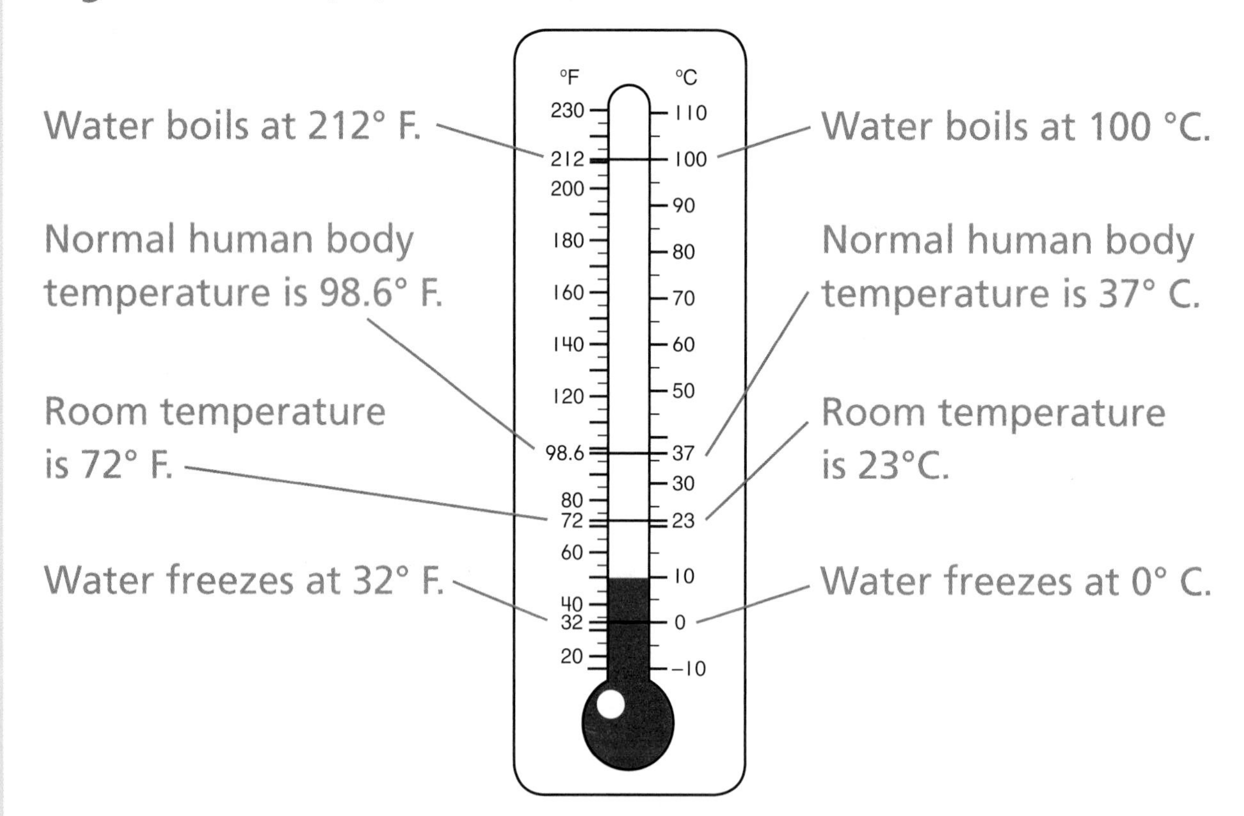

Read the thermometer. Write the temperature in °F and in °C.

50 °F = 10 °C

This is a different kind of thermometer.

It shows the temperature ________ °F.

 Read each thermometer. Write the temperature in °F and in °C.

1 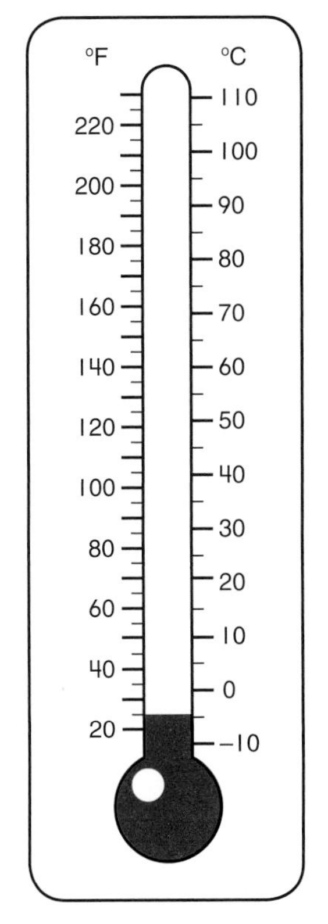

_________ °F = _________ °C

2 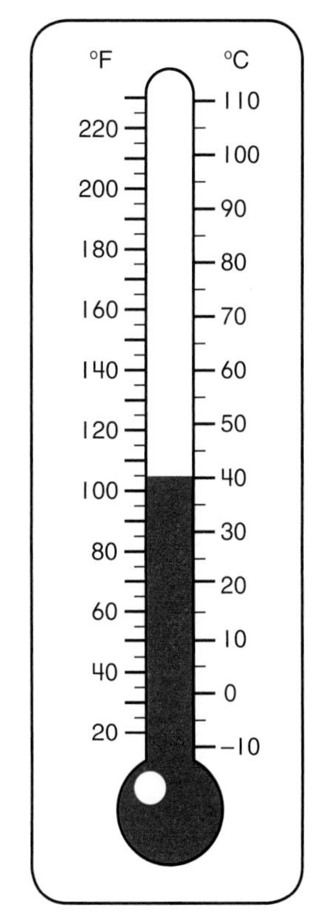

_________ °F = _________ °C

 Read each thermometer. Write each temperature. Circle °F or °C.

3 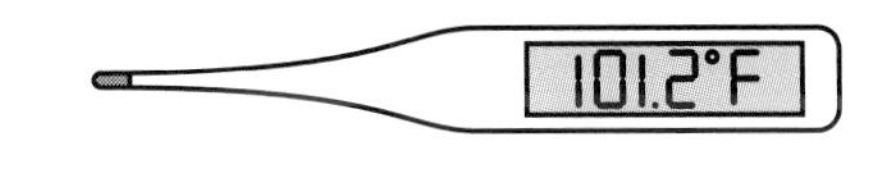

_________ °C or °F

4 

_________ °C or °F

 Write each set of temperatures in order from lowest to highest. You can use a thermometer from above to help you.

5 40°C, 90°F, 53°C _________, _________, _________

6 12°F, 78°F, 0°F _________, _________, _________

7 105°F, 95°F, 60°C _________, _________, _________

8 95°C, 47°C, 62°C _________, _________, _________

Lesson 44

Explore Time

There are 60 minutes in one hour. There are 24 hours in 1 day.
You can read time in different ways.

- 6:00
- six o'clock

- 1:45
- quarter to 2
- forty-five minutes after 1

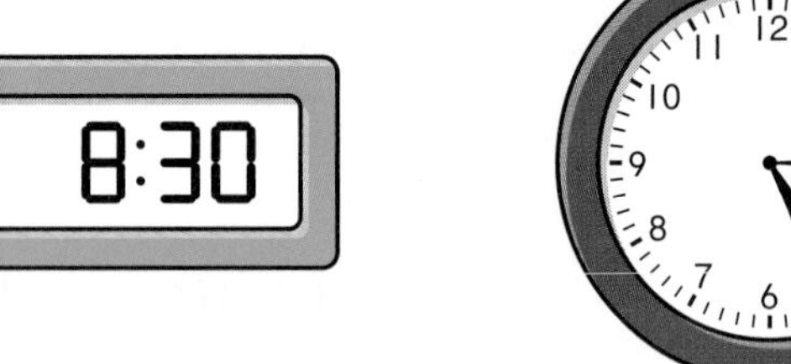

- 8:30
- thirty minutes after 8
- half past 8

- 5:15
- fifteen minutes after 5
- quarter past five

▶ Read the first clock. Write the time on the second clock.

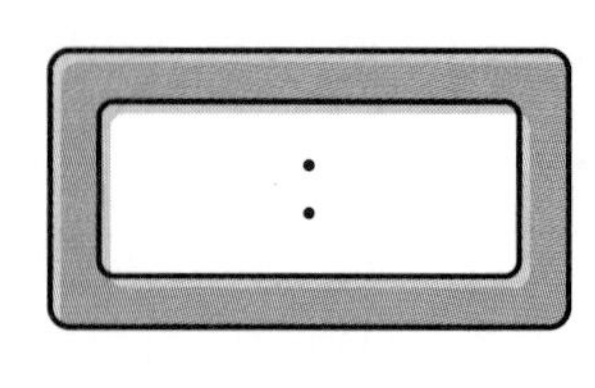

There are 5 minutes between each number on the clock.
The clock shows 25 minutes after 11. Write 11:25.

Connect the clocks that show the same time.

1

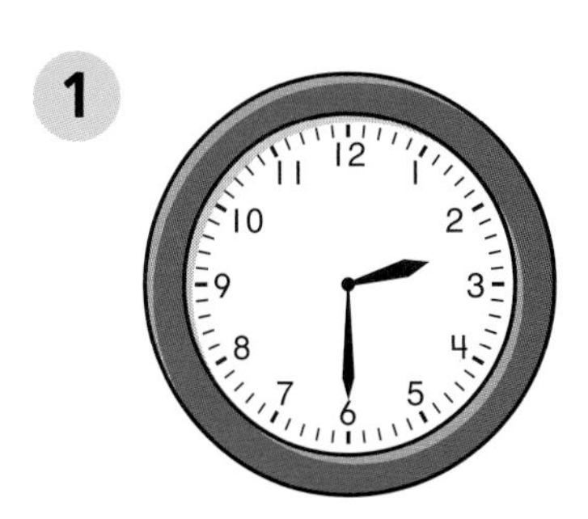

2

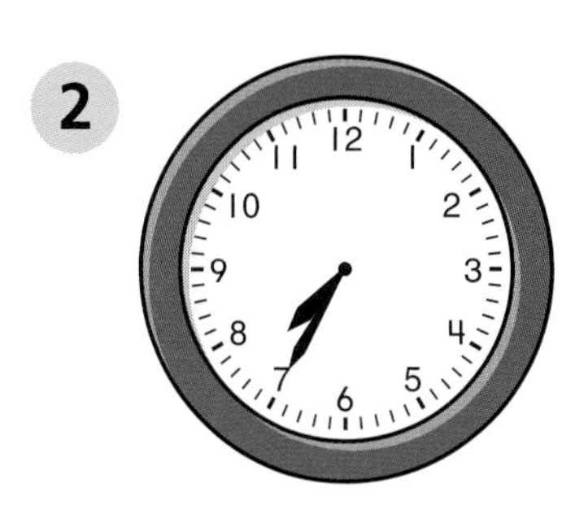

3

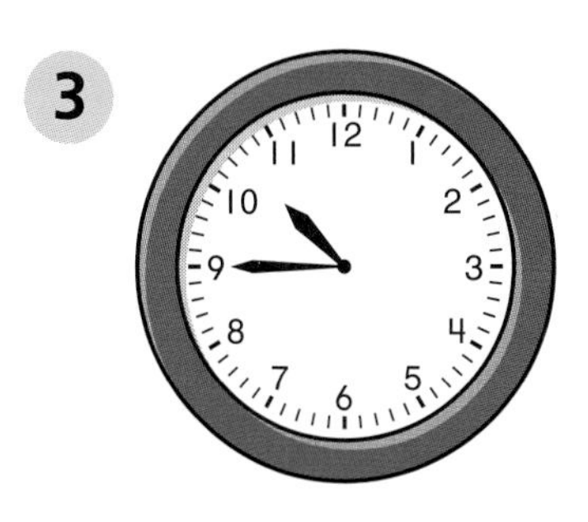

Read the clock. Write the time.

4

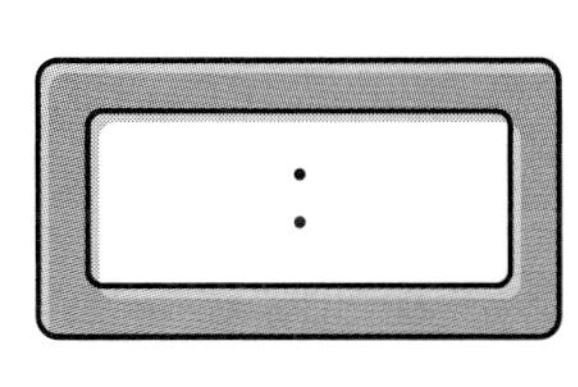

5

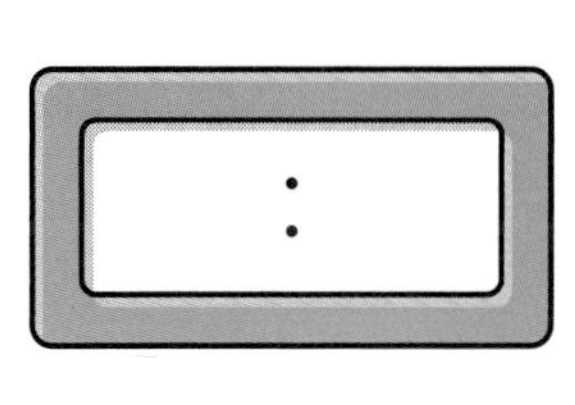

6

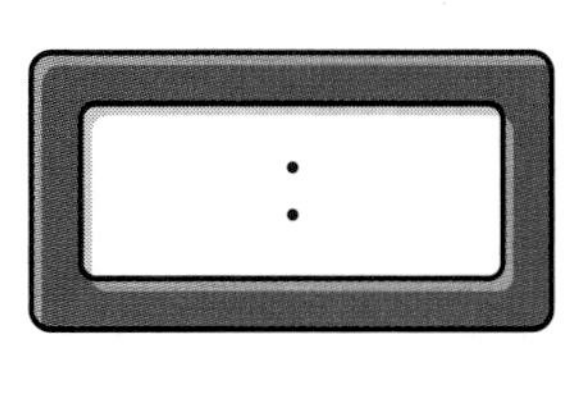

7

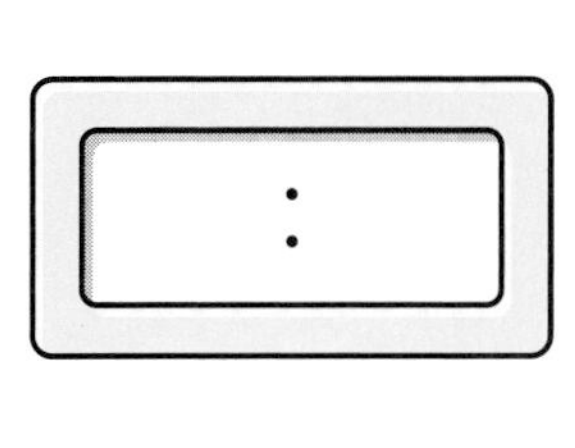

Checkup

Circle the weight or mass of each object.

1

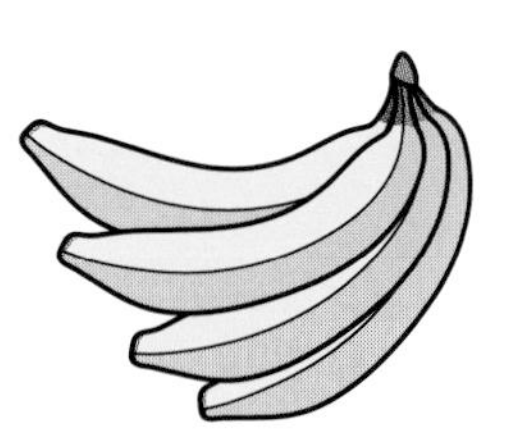

about 2 ounces | about 2 pounds

2

about 12 ounces | about 12 pounds

3

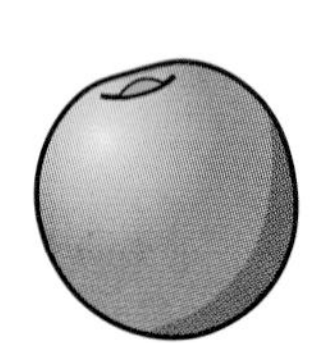

about 80 grams | about 80 kilograms

4

about 5 grams | about 5 kilograms

Circle the unit you would use to measure the weight or mass of each object.

5 watermelon 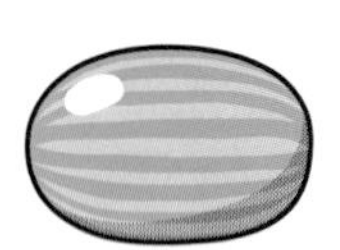 ounces or pounds

6 pencil 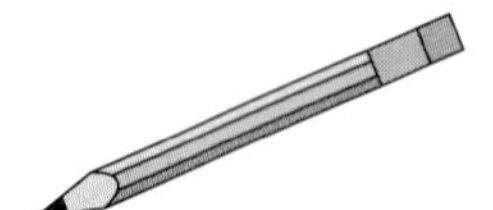 grams or kilograms

Fill in the blank of the missing measure.

7 1 lb = _______ oz

8 32 oz = _______ lb

9 2 kg = _______ g

10 3,000 g = _______ kg

Circle the mass of each object.

11

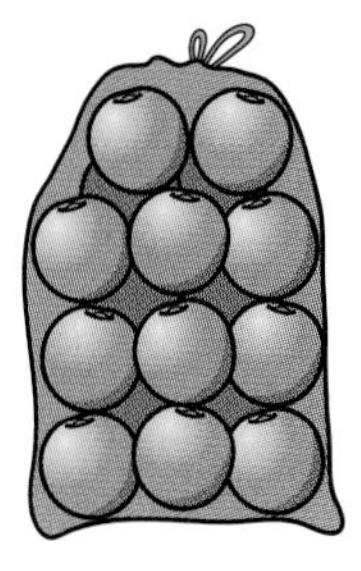

about 3 grams | about 3 kilograms

12

about 5 grams | about 5 kilograms

Write each temperature.

13

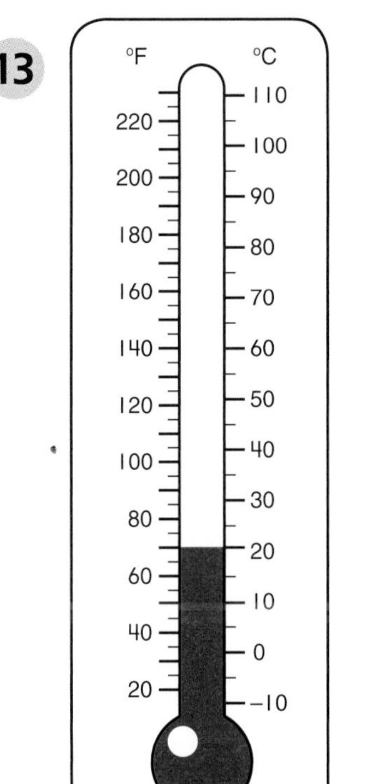

_______ °F = _______ °C

14

_______ °F

15

100.0°C

_______ °C

Read the clock. Write the time.

16

17

Lesson 45

Explore Probability

Probability describes the chance that something will happen. Whatever happens is called an **outcome**.

Use the spinner to answer each question.

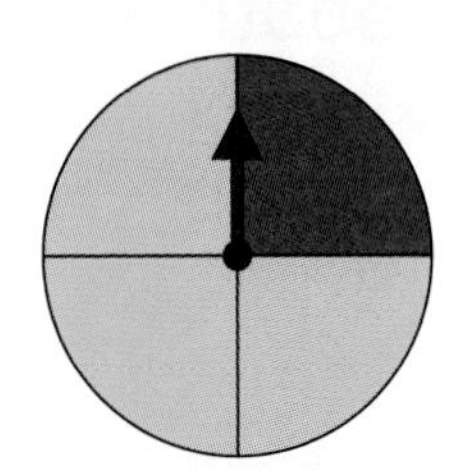

How many colors are on the spinner? ______

How many possible outcomes are there? ______

Fill in the blanks.

The spinner has ______ parts in all. 3 out of 4 parts are yellow.

So, there are 3 out of 4 chances that the pointer will stop on yellow. ______ of the parts is purple. So, there is a ______ out of ______ chance that the pointer will stop on purple.

Fill in the blanks.

1 Imagine tossing a penny.

The penny has ______ sides.

There are ______ possible outcomes.

2 A bag holds 3 green, 2 yellow, and 4 blue marbles. Imagine taking one marble from the bag.

There are ______ different colors of marbles.

There are ______ possible outcomes.

3 Imagine spinning the spinner.

There are ______ different symbols on the spinner.

There are ______ possible outcomes.

Write the probability.

4 Imagine you toss a nickel. The nickel lands heads-up.

The nickel has ______ sides.

______ of the sides is heads. The other side is tails. There is a ______ out of ______ chance that the nickel will land heads-up.

5 A bag holds 2 green, 3 yellow, and 5 blue marbles. Imagine taking a green marble from the bag.

There are ______ marbles in the bag.

______ of the marbles are green.

There is a ______ out of ______ chance of taking a green marble.

Describing Probability

▶ **Use the words from the box to tell how likely it is that the pointer will stop on orange after a spin.**

less likely	more likely	equally likely

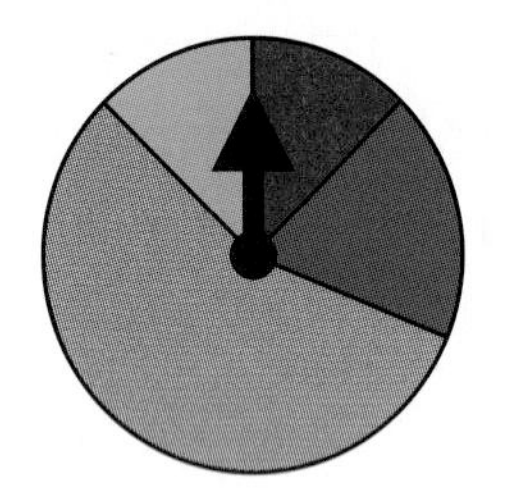

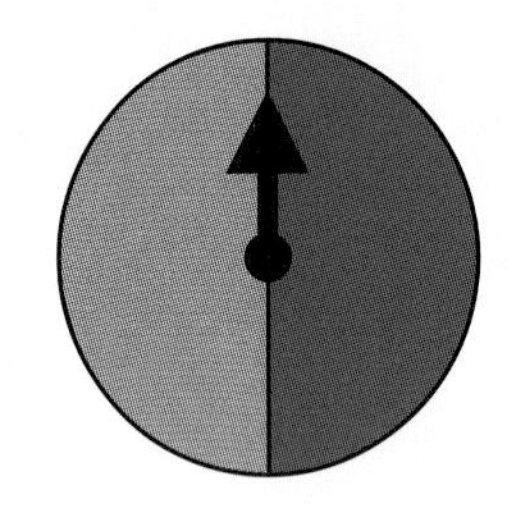

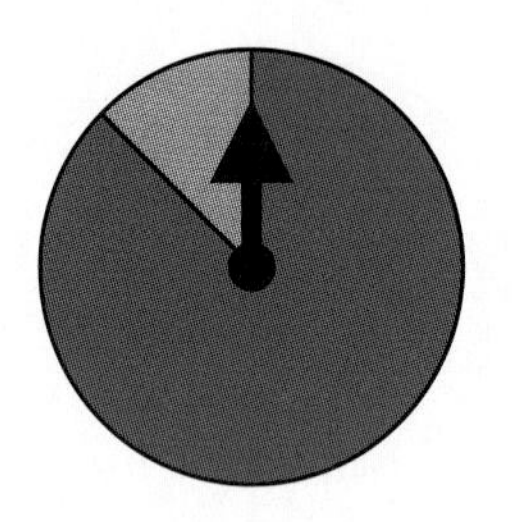

A. ____________ **B.** ____________ **C.** ____________

▶ **Circle the word that best describes the probability of taking a green marble from each jar without looking.**

D. certain

impossible

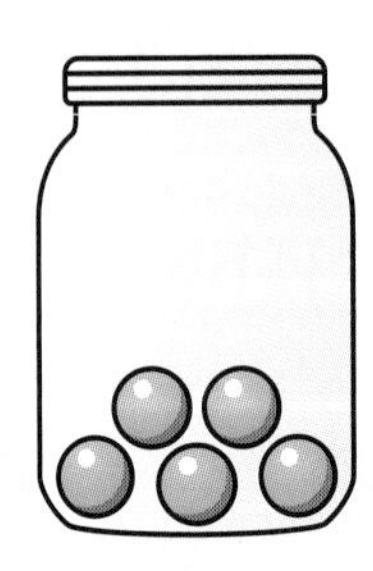

E. likely

unlikely

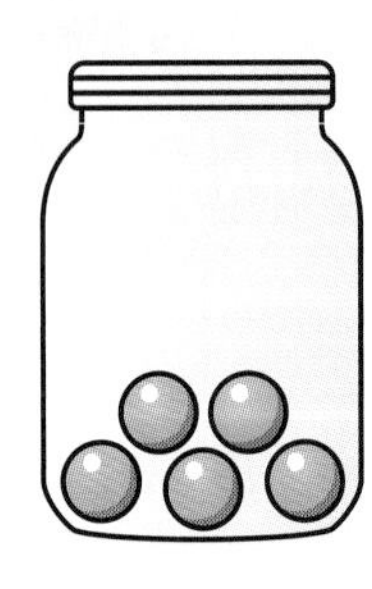

▶ **Circle the word that best describes the probability that the arrow on each spinner will stop on 5.**

1 likely

unlikely

2 certain

impossible

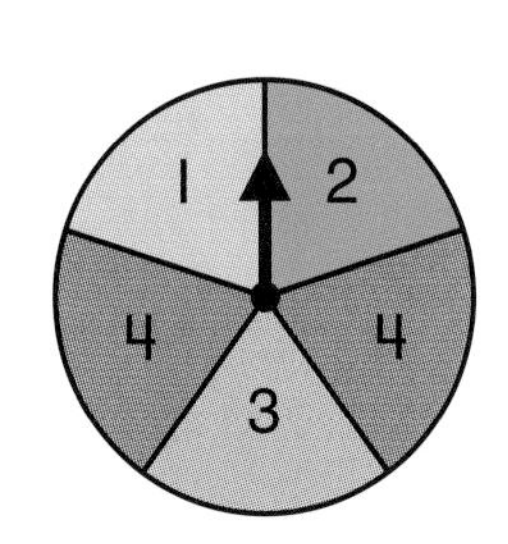

 **Imagine closing your eyes and taking a shape out of each box. Write words to tell how likely it is that you will take a ☆ from each box.**

3 

____

4 

____

5 

____

Draw lines to match each term to its best example.

6 equally likely

7 more likely

8 less likely

9 impossible

10 certain

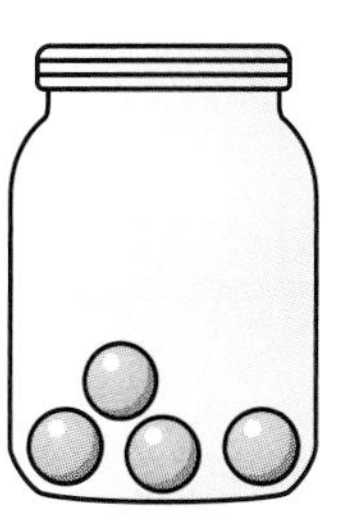

picking a yellow marble from this jar

a coin landing heads-up after a toss

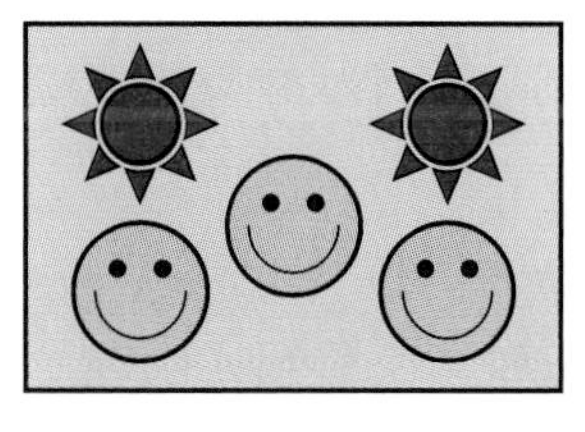

taking a face from the box

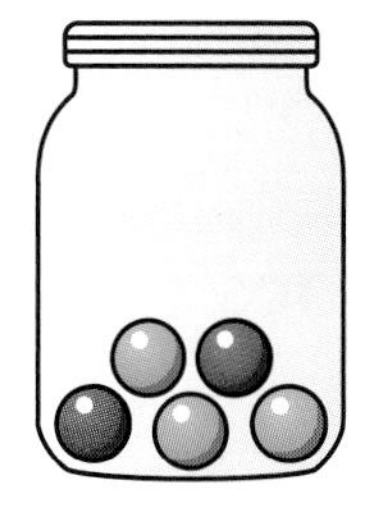

taking a red marble from this jar

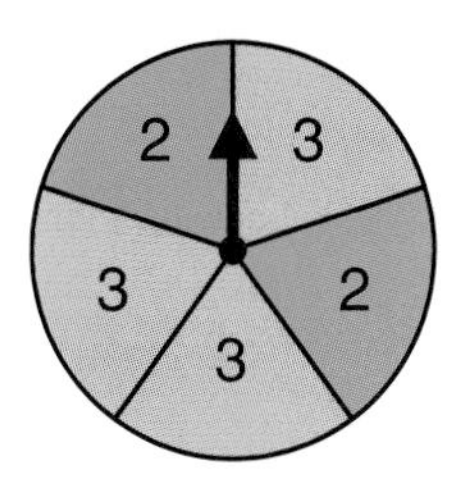

the pointer stopping on 1 after a spin

Checkup

1 Imagine tossing a dime.

The dime has ______ sides.

There are ______ possible outcomes.

2 Imagine choosing a shape from the box.

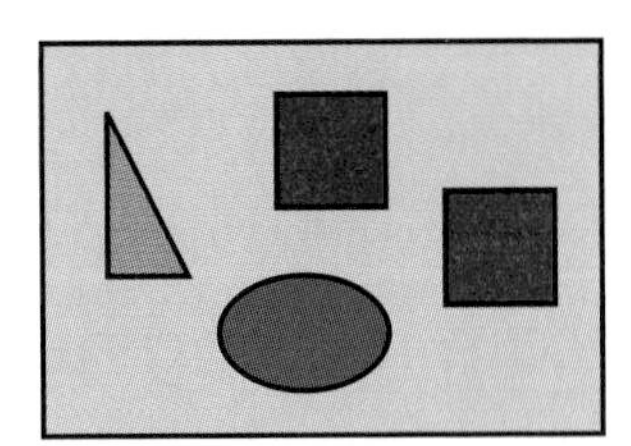

There are ______ different kinds of shapes.

There are ______ possible outcomes.

Fill in the blanks.

3

The nickel has ______ sides. ______ of the sides is tails.

There is a ______ out of ______ chance that the coin will land tails-up.

How likely is it that the pointer will stop on 1 after a spin? Choose a term from the box to write your answer.

certain	impossible
equally likely	more likely	less likely

4 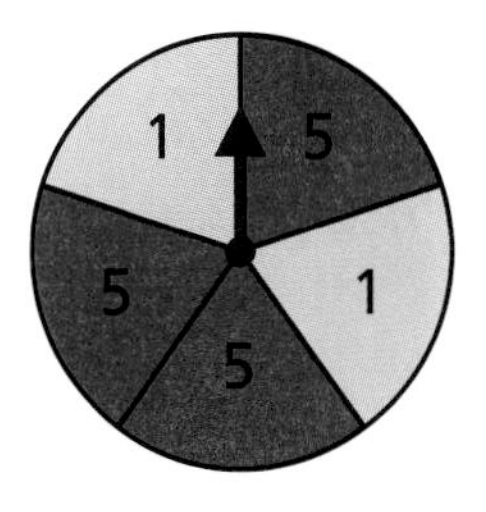

____

5 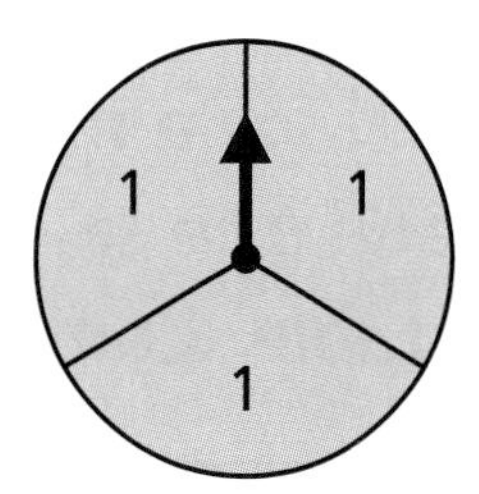

____

Draw lines to match each term to its best example.

6 more likely

taking a moon symbol from the box

7 certain

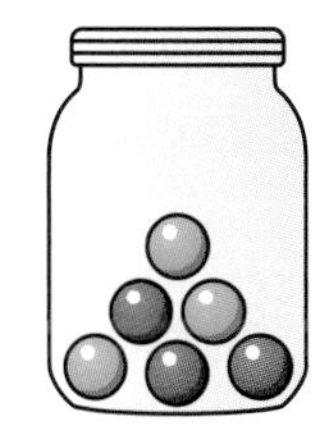

taking a red marble from this jar

8 less likely

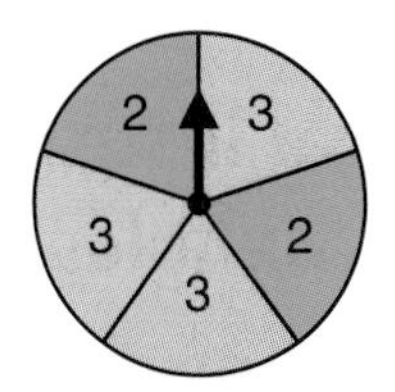

the pointer stopping on 3 after a spin

9 equally likely

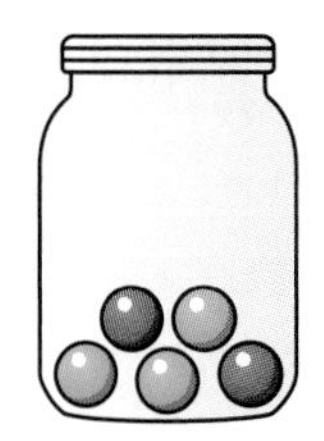

taking a green marble from this jar

10 impossible

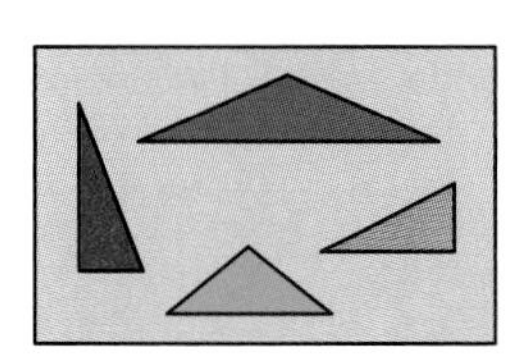

choosing a triangle from the box

Frequency Tables

You can collect data, or information. Then you can organize the data in a table. Look at this example.

A class voted for a name for their new pet hamster. The teacher made one tally mark for each student's vote in a frequency table. Complete the table.

Names	Tally	Frequency
Bear	IIII	4
Woolly Mammoth	~~IIII~~ II	
Speedy	~~IIII~~	
Twitch	III	

How many students voted? ______

What name did most students vote for? ______________________

A music teacher gave musical instruments to his students. 4 students got bells, 9 got triangles, 5 got xylophones, and 7 got rhythm sticks. Put the data in this frequency table.

Musical Instruments	Tally	Frequency
Bells		
Triangles		
Xylophones		
Rhythm sticks		

1 How many students got triangles? ______

2 How many students got musical instruments? ______

A gym teacher asked her class to vote for their favorite activity. She put the results in this frequency table. Complete the table.

Activity	Tally	Frequency
Gymnastics	𝍸 𝍸 I	
Track	𝍸 IIII	
Volleyball	IIII	

3 How many students voted for track? ______

4 How many students voted in all? ______

Some students voted for their favorite kind of transportation and put the data in this frequency table. Complete the table.

Transportation	Tally	Frequency
Train	𝍸 II	
Ship	III	
Airplane	II	
Hot-air balloon	𝍸 𝍸 II	

5 Students' most favorite kind of transportation is ______________.

6 Their least favorite kind of transportation is ______________.

Line Plots

You can put the data in a frequency table into a line plot.

▶ **Use the frequency table to complete the line plot.**

Pets People Own		
Number of Pets	**Tally**	**Frequency**
0	II	2
1	III	3
2	~~IIII~~	5
3	IIII	4
4	II	2

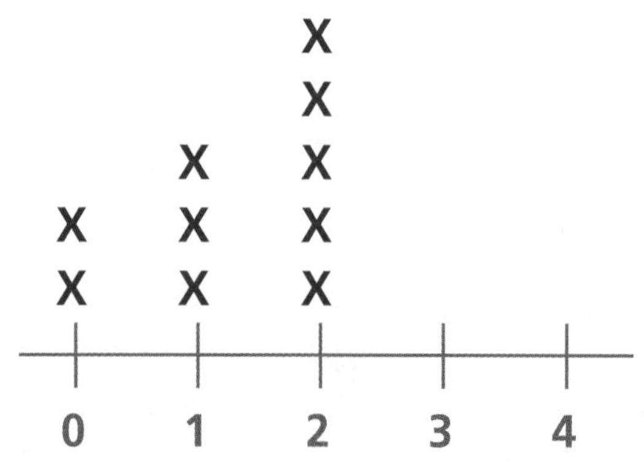

How many students own 1 pet? ______

What is the greatest number of pets people own? ______

What is the fewest number of pets people own? ______

1 Use the frequency table to complete the line plot.

Club Meetings After School		
Day	**Tally**	**Frequency**
Monday	III	3
Tuesday	IIII	4
Wednesday	IIII	4
Thursday	~~IIII~~	5
Friday	I	1

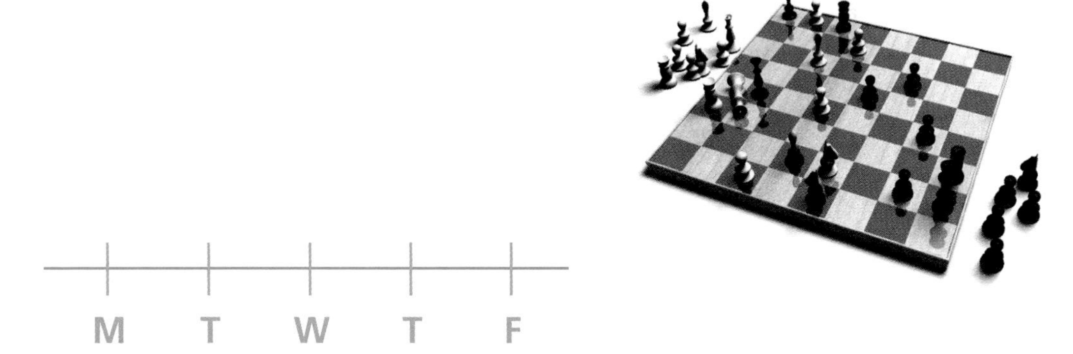

Number of Club Meetings After School Each Week

2 How many clubs meet on Wednesdays after school? ______

3 On which day do the most clubs meet? ____________________

4 On which day do the fewest clubs meet? ____________________

5 On which days do the same number of clubs meet?

____________________ and ____________________

6 How many clubs meet after school each week? ______________

Pictographs

A pictograph uses symbols to stand for data. The key tells you how many each symbol stands for.

Students wrote a class story about an imaginary superhero mouse. They voted for the mouse's name. Use the frequency table to complete the pictograph.

Favorite Superhero Names		
Superhero Names	**Tally**	**Frequency**
Gray Tail	II	2
Tip Toe	𝍸 I	6
Supermouse	IIII	4
Swifty	𝍸 III	8

Favorite Superhero Names	
Gray Tail	
Tip Toe	
Supermouse	
Swifty	

Which name got the most votes?

Swifty

Which name got the least votes?

____

How many votes does each symbol stand for? ______

1 Students in the Nature Club voted for their favorite backyard birds. Put the data in the frequency table into the pictograph.

Favorite Backyard Birds		
Birds	**Tally**	**Frequency**
Sparrow	III	3
Bluejay	~~IIII~~	5
Cardinal	~~IIII~~ II	7
Mockingbird	II	2

Favorite Backyard Birds	
Sparrow	
Bluejay	
Cardinal	
Mockingbird	

= 1 bird

2 What bird did students vote for the **most**? ____________________

3 What bird did students vote for the **least**? ____________________

4 How many more votes did cardinals get than bluejays? ______

5 How many students voted? ______

Bar Graphs

An art teacher put the results of a contest in a frequency table. Use the frequency table to complete the bar graph.

Ribbons Won		
Colors	**Tally**	**Frequency**
Blue	////	4
Red	~~////~~ //	7
Yellow	////	4

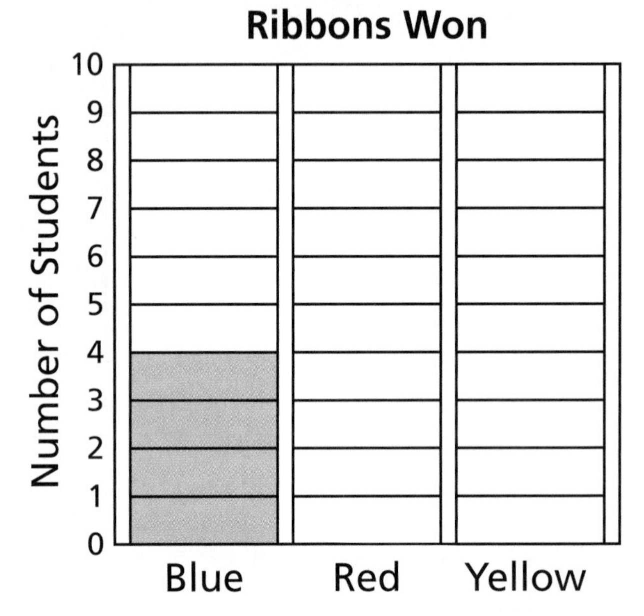

To find the mean, add all the values. Then divide by the number of values. 4 + 7 + 4 = 15. 15 ÷ 3 = 5. The mean is ______.

To find the median, order the values from least to greatest.

The median is the middle value of the ordered set. 4, 4, 7

The median is ______.

To find the mode, find the value that occurs most often in the ordered set. 4, 4, 7 The mode is ______.

To find the range, subtract the least value from the greatest value.

7 − 4 = ______ The range is ______.

1 Students in the Music Club voted for a color for their new T-shirts. Put the data in the frequency table into a bar graph.

Color Choice		
Colors	**Tally**	**Frequency**
White (W)	II	2
Black (B)	II	2
Orange (O)	IIII	4
Purple (P)	IIII	4
Yellow (Y)	III	3

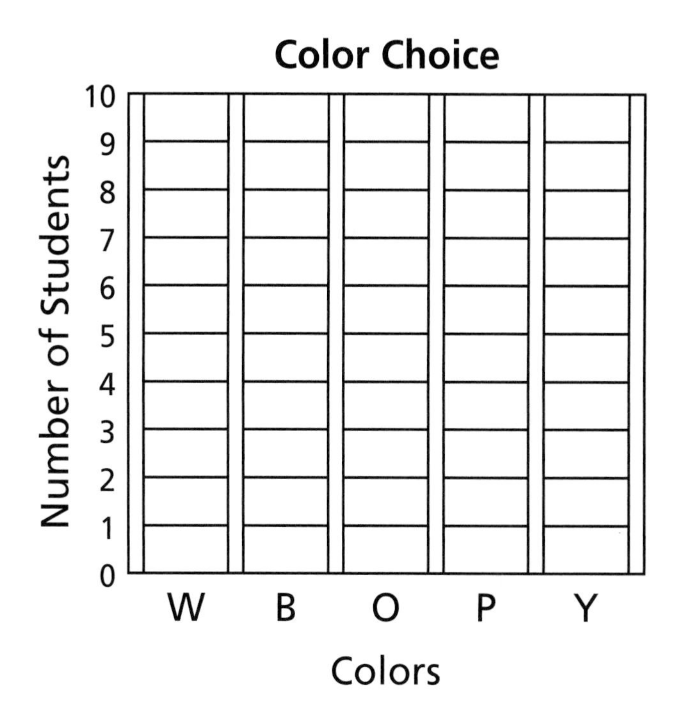

2 Find the mean. 2 + 2 + 4 + 4 + 3 = ______

______ ÷ ______ = ______ The mean is ______.

3 Find the median. 2, 2, 3, 4, 4

The median is ______.

4 Find the mode. 2, 2, 3, 4, 4

There are two modes. They are ______ and ______.

5 Find the range. ______ − ______ = ______

The range is ______.

Checkup

1 A teacher asked his class to vote for their favorite free-time activity He put the results in this frequency table. Complete the table.

Favorite Free-Time Activities		
Activity	**Tally**	**Frequency**
Reading	HHT II	
Sports	IIII	
Friends	HHT II	
Hobbies	HHT IIII	

2 How many students voted for playing with friends?

____

3 How many students voted in all?

____

4 The results of a school track meet are in the following frequency table. Use the data in the table to make a line plot.

Track Meet Results		
Number of Medals Won	**Tally**	**Frequency**
0	II	2
1	III	3
2	HHT	5
3	I	1

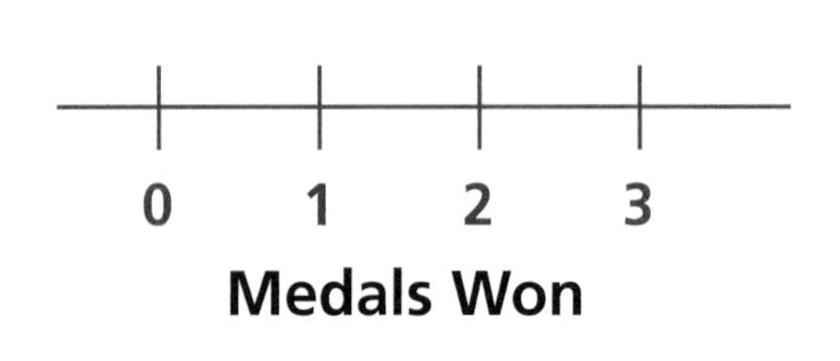

5 How many students won 3 medals? _______

6 What number of medals were won the most? _______

7 A group of art students voted on what they like to draw most. Use the data in the frequency table to make a pictograph.

What We Like to Draw		
Subject	**Tally**	**Frequency**
Plants	卌	5
Animals	卌 III	8
People	卌 I	6
Land	卌 II	7
Oceans	IIII	4

What We Like to Draw	

= 1 person

8 What did students vote for least often? ______________________

9 What did students vote for most often? ______________________

10 A coach wrote the number of laps students jogged in one day. Put the data in this frequency table into the bar graph.

Laps Students Jogged		
Laps	**Tally**	**Frequency**
1	𝍸	5
2	𝍸 II	7
3	𝍸 II	7
4		0
5	I	1

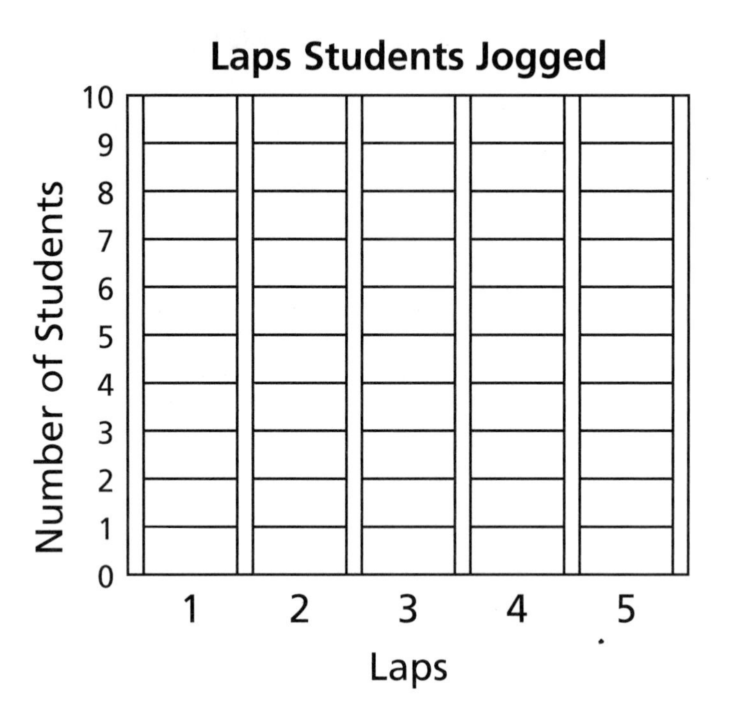

11 Find the mean. $0 + 1 + 5 + 7 + 7 =$ ______

______ $\div 5 =$ ______ The mean is ______.

12 Find the median. ______, ______, ______, ______, ______.

The median is ______.

13 Find the mode. 0, 1, 5, 7, 7 The mode is ______.

14 Find the range. ______ $-$ ______ $=$ ______

The range is ______.